139
Structure and Bonding

Series Editor: D. M. P. Mingos

For further volumes:
http://www.springer.com/series/430

Structure and Bonding

Series Editor: D. M. P. Mingos

Recently Published and Forthcoming Volumes

Zintl Ions
Principles and Recent Developments
Volume Editor: Thomas F. Fässler
Vol. 140, 2011

Zintl Phases
Principles and Recent Developments
Volume Editor: Thomas F. Fässler
Vol. 139, 2011

Inorganic 3D Structures
Volume Editor: Angel Vegas
Vol. 138, 2011

Molecular Catalysis of Rare-Earth Elements
Volume Editor: Peter W. Roesky
Vol. 137, 2010

Metal-Metal Bonding
Volume Editor: Gerard Parkin
Vol. 136, 2010

Functional Phthalocyanine Molecular Materials
Volume Editor: Jianzhuang Jiang
Vol. 135, 2010

Data Mining in Crystallography
Volume Editors: Hofmann, D. W. M., Kuleshova, L. N.
Vol. 134, 2010

Controlled Assembly and Modification of Inorganic Systems
Volume Editor: Wu, X.- T.
Vol. 133, 2009

Molecular Networks
Volume Editor: Hosseini, M. W.
Vol. 132, 2009

Molecular Thermodynamics of Complex Systems
Volume Editors: Lu, X., Hu, Y.
Vol. 131, 2009

Contemporary Metal Boron Chemistry I
Volume Editors: Marder, T. B., Lin, Z.
Vol. 130, 2008

Recognition of Anions
Volume Editor: Vilar, R.
Vol. 129, 2008

Liquid Crystalline Functional Assemblies and Their Supramolecular Structures
Volume Editor: Kato, T.
Vol. 128, 2008

Organometallic and Coordination Chemistry of the Actinides
Volume Editor: Albrecht-Schmitt, T. E.
Vol. 127, 2008

Halogen Bonding
Fundamentals and Applications
Volume Editors: Metrangolo, P., Resnati, G.
Vol. 126, 2008

High Energy Density Materials
Volume Editor: Klapötke, T. H.
Vol. 125, 2007

Ferro- and Antiferroelectricity
Volume Editors: Dalal, N. S., Bussmann-Holder, A.
Vol. 124, 2007

Photofunctional Transition Metal Complexes
Volume Editor: V. W.W. Yam
Vol. 123, 2007

Single-Molecule Magnets and Related Phenomena
Volume Editor: Winpenny, R.
Vol. 122, 2006

Zintl Phases

Principles and Recent Developments

Volume Editor: Thomas F. Fässler

With contributions by

J. Evers · U. Häussermann · K. Kovnir · V.F. Kranak ·
G.J. Miller · K. Puhakainen · M.W. Schmidt ·
A.V. Shevelkov · F. Wang · T.-S. You

Springer

Editor
Thomas F. Fässler
Department of Chemistry
Technical University Munich
Lichtenbergstraße 4
85747 Garching
Germany
thomas.faessler@lrz.tum.de

ISSN 0081-5993 e-ISSN 1616-8550
ISBN 978-3-642-21149-2 e-ISBN 978-3-642-21150-8
DOI 10.1007/978-3-642-21150-8
Springer Heidelberg Dordrecht London New York

Library of Congress Control Number: 2011929694

Cover design: eStudio Calamar, Berlin/Figueres, Spain

Printed on acid-free paper

Springer is part of Springer Science+Business Media (www.springer.com)

Structure and Bonding
Also Available Electronically

Structure and Bonding is included in Springer's eBook package *Chemistry and Materials Science*. If a library does not opt for the whole package the book series may be bought on a subscription basis. Also, all back volumes are available electronically.

For all customers who have a standing order to the print version of *Structure and Bonding*, we offer the electronic version via SpringerLink free of charge.

If you do not have access, you can still view the table of contents of each volume and the abstract of each article by going to the SpringerLink homepage, clicking on "Chemistry and Materials Science," under Subject Collection, then "Book Series," under Content Type and finally by selecting *Structure and Bonding*.

You will find information about the

- Editorial Board
- Aims and Scope
- Instructions for Authors
- Sample Contribution

at springer.com using the search function by typing in *Structure and Bonding*.

Color figures are published in full color in the electronic version on SpringerLink.

Aims and Scope

The series *Structure and Bonding* publishes critical reviews on topics of research concerned with chemical structure and bonding. The scope of the series spans the entire Periodic Table and addresses structure and bonding issues associated with all of the elements. It also focuses attention on new and developing areas of modern structural and theoretical chemistry such as nanostructures, molecular electronics, designed molecular solids, surfaces, metal clusters and supramolecular structures. Physical and spectroscopic techniques used to determine, examine and model structures fall within the purview of *Structure and Bonding* to the extent that the focus is on the scientific results obtained and not on specialist information concerning the techniques themselves. Issues associated with the development of bonding models and generalizations that illuminate the reactivity pathways and rates of chemical processes are also relevant.

The individual volumes in the series are thematic. The goal of each volume is to give the reader, whether at a university or in industry, a comprehensive overview of an area where new insights are emerging that are of interest to a larger scientific audience. Thus each review within the volume critically surveys one aspect of that topic and places it within the context of the volume as a whole. The most significant developments of the last 5 to 10 years should be presented using selected examples to illustrate the principles discussed. A description of the physical basis of the experimental techniques that have been used to provide the primary data may also be appropriate, if it has not been covered in detail elsewhere. The coverage need not be exhaustive in data, but should rather be conceptual, concentrating on the new principles being developed that will allow the reader, who is not a specialist in the area covered, to understand the data presented. Discussion of possible future research directions in the area is welcomed.

Review articles for the individual volumes are invited by the volume editors.

In references *Structure and Bonding* is abbreviated *Struct Bond* and is cited as a journal.

Impact Factor in 2009: 4.152; Section "Chemistry, Inorganic & Nuclear": Rank 2 of 43; Section "Chemistry, Physical": Rank 7 of 113

Preface

"Damit aber wird alle Grundlagenforschung zur Zweckforschung auf weite Sicht" [1]. This important statement given in a lecture on *Intermetallic Compounds* at a conference of the German Chemical Society in 1932 indicated the visionary power of the German chemist Eduard Zintl (1898–1941), since the subclass of intermetallic compounds, which he had discovered, and which have been named "Zintl phases" by *Laves* in 1941 [2], advanced to one of most inventory subjects in materials science during the last century.

During the ensuing period, Zintl's original understanding of these polar intermetallic compounds as salt-like species, in which the valence electrons are formally transferred from the more electropositive to the more electronegative component, has relocated the unadulterated description of intermetallic compounds in terms of geometrical aspects such as coordination numbers, coordination polyhedra and atomic radii ratios, and their electronic situations which had been considered as delocalized electronic systems by means of free-electron gas models and average electron numbers (valence electron concentration). This new approach has led to the idea that metal atoms in Zintl phases form localized covalent and even multiple and multi-center chemical bonds solely based on the valence rules and molecular orbital descriptions as generally accepted in molecular chemistry. At the same time, it has helped to predict their structures and properties and thus has made Zintl phases most valuable candidates for the tuning of specific structures and electronic/magnetic properties. In the case of soluble Zintl phases resulting solutions contain so-called *Zintl anions*. This important field is treated within this series in Volume 140 *Zintl Ions* (edited by *T.F. Fässler*).

In this book, the progress in the field of Zintl phases from Zintl's original ideas to the application of these compounds in materials science is outlined. In the first chapter, *GJ Miller et al.* report on quantitative advances in the Zintl–Klemm formalism based on first-principle calculations. The next chapter by *J Evers* is dedicated to the behavior of Zintl phases under high pressure and deals with *in situ* experimental studies complemented by theoretical investigations on the basis of DFT calculations. The very rich chemistry of Zintl phases with clathrate-type structures, in which 40% of all stable elements of the Periodic Table are involved, is presented by *A Shevelkov and K Kovnir* in another chapter. The reader is also directed to the chapter of Santamaría-Pérez and Liebau in Volume 138 of Structure and Bonding edited by *A Vegas*, which shows in detail the structural relationship of Zintl-type clathrates, porous tectosilicates, and clathrate hydrates. In the last chapter, *U Häussermann et al.* describe the reactions of Zintl phases with elemental hydrogen which can lead to interstitial hydrides as well as to polyanionic hydrides with hydrogen covalently bound to atoms of the polyanions of a Zintl phase, and thus makes hydrogen-containing Zintl phases important model systems for fundamental studies of hydrogen–metal interactions.

Reopening ceremony of the Eduard Zintl Institute in 2002 at the Technical University of Darmstadt (from left to right: TF Fässler (director), KH Lieser, bust of Eduard Zintl, B Eisenmann, H-F Klein, J Brickmann).

[1] Zintl E (1939) Intermetallische Verbindungen. Angew Chem 52:1 (*Thus, all fundamental research becomes applied research on a long term*)
[2] Laves F (1941) Naturwissenschaften 29:244 (*Eduard Zintls Arbeiten über die Chemie und Struktur von Legierungen*)

Technical University Munich
April 2011

T.F. Fässler

Contents

Struct Bond (2011) 139: 1–55
DOI:10.1007/430_2010_24

Published online: 27 October 2010

Quantitative Advances in the Zintl–Klemm Formalism

Gordon J. Miller, Michael W. Schmidt, Fei Wang, and Tae-Soo You

Abstract The Zintl–Klemm formalism has enjoyed tremendous success for rationalizing numerous network- and cluster-based structures involving main group elements. As research continues to explore the applicability of this potentially predictive concept, developments in theoretical and computational chemistry allow the study of larger and heavier molecular and solid-state building blocks to test this powerful formalism semiquantitatively, as well as improved handling of interatomic interactions involving widely disparate elements. Inherent in the Zintl–Klemm formalism is a coexisting tension between anisotropic, covalent bonding interactions, and isotropic, ionic, or metallic bonding forces collected in a system whose equilibrium volume is governed by atomic sizes via core repulsions. This chapter summarizes recent applications and quantitative developments of the Zintl–Klemm formalism, emphasizing results of first-principles calculations on molecules and extended solids, as well as selected experimental results that address the general validity of using this simple concept.

Keywords Electronic Structure · Zintl-Klemm Formalism

Contents

G.J. Miller (✉), M.W. Schmidt, F. Wang
Department of Chemistry, Iowa State University and Ames Laboratory, US-DOE, 321, Spedding Hall, Ames, IA 50011-3111, USA
e-mail: gmiller@iastate.edu
T.-S. You
Department of Chemistry, Chungbuk National University, 410 Sungbong-ro, Heungduku, Cheongju, Chungbuk, South Korea, 361-763

1 Introduction

Chemists are particularly enamored with valence electron counting rules because they can be useful for rationalizing as well as predicting the chemical behavior of substances, i.e., structures and reactivities. Some of the more common electron counting rules include the *octet* (*Lewis–Langmuir*) *rule* [1], which is broadly applied to problems in organic, main-group inorganic, and biochemistry; the *18-electron rule* [2], valuable for organometallic complexes; the *valence shell electron pair repulsion* (*VSEPR*) *theory* [3, 4]; and the *Wade–Mingos rules* [5, 6] for deltahedral clusters commonly observed in boranes, carboranes, and metallocarboranes. The inherent simplicity of valence rules relies heavily on the Periodic Table of the Elements, which has its foundations in the ground states and configurations of each element, and the successes of the quantum theory applied to molecules and extended solids [7]. Of course, when electron counting rules work, their predictive role tends to expand because our confidence level in their reliability increases. On the other hand, when they do not work, chemists begin to ponder their general validity and question the earlier interpretations. That is the nature of the scientific method.

The *Zintl–Klemm concept* is another such valence electron counting scheme that has achieved wide usage among molecular and solid-state disciplines [8–11]. At the time of Eduard Zintl's research [12–14], which included the study of intermetallics involving combinations of active with noble metals or posttransition elements in liquid ammonia, Hume-Rothery and Westgren had already presented their insightful correlation between structure and valence electron concentration for noble metal intermetallics [15–17], while the Grimm–Sommerfeld rule had just been proposed for tetrahedral frameworks [18]. Due to some of the unusual properties of the intermetallics studied by Zintl, properties that included high melting points, high heats of formation, poor electrical conductivities notwithstanding their metallic luster, and increased brittleness, Zintl concluded that the network adopted by the noble metal arose from charge (electron) transfer from the active metal. For the traditional example of a *Zintl phase*, NaTl, Zintl concluded that the diamond network adopted by the Tl atoms arises when each Na atom donates its valence electron to Tl [19]. In this way, both the arrangement of Na and Tl atoms in NaTl as well as the compressed spatial requirements of the Na atoms within the three-dimensional framework of Tl atoms could be rationalized. Thus, as Schäfer pointed out [10], "ionic and covalent bonds in intermetallic phases were a matter of course for Zintl". Zintl's conjecture was also consistent with Pauling's assessment of elemental electronegativities [20]. Klemm generalized this powerful idea for such compounds [21, 22], formulated as $A_aX_x \wedge= (A^{n+})_a[X^{(an/x)-}]_x$, A = active metal and X = noble (electronegative) metal or semimetal, by proposing the *pseudoatom concept* for the electronegative components, which means these elements would exhibit a structure based on nearest-neighbor atomic connectivities, characteristic of the isovalent element. In this way, NaTl can be formulated as $Na^{+}Tl^{-}$, in which the new tetravalent Tl^{-} forms a three-dimensional, quasi-infinite diamond network, $^{3}_{\infty}[Tl]^{-}$, as observed for the ground state structures of Si, Ge, and Sn, but not Pb!

Compounds following this Zintl–Klemm formalism are called, in general, *Zintl phases*, which can range from extended solids to cluster compounds, i.e., containing cluster anions, to salt-like structures, as in Mg_2Sn [8, 9, 23, 24]. To separate Zintl phases from intermetallics and insulators, Nesper defines three criteria for Zintl phases [11]: (1) a well-defined relationship exists between their chemical (geometrical) and electronic structures (i.e., certain aspects of their structures satisfy electron counting rules); (2) they are semiconductors (energy gaps less than 2 eV), or, at least, show increasing electrical conductivities with increasing temperature; and (3) they are mostly diamagnetic, but, if paramagnetic, they should not show temperature-independent (Pauli) paramagnetism. These criteria imply that Zintl phases (like valence compounds) have narrow homogeneity widths (i.e., they are mostly line compounds) and could display semimetallic behavior. In addition to the "observable" criteria, the calculated electronic structures of Zintl phases typically show that the bonding and nonbonding states are completely occupied and separated from the empty, antibonding states by not more than 2 eV. As the electronegative component progresses from the right to the left of the Zintl border, this energy gap approaches zero, but the chemical system retains important characteristics of Zintl phases, e.g., "metallic Zintl phases" [25].

Since Zintl phases provide structural and electronic connections among the traditional classes of *metallic*, *covalent*, and *ionic* compounds, these three classifications can be demonstrated using a van Arkel–Ketelaar map, illustrated in Fig. 1 [28, 29]. This diagram is constructed quantitatively, using as the horizontal coordinate the sum of configuration energies [26, 27], and as the vertical axis, their difference for binary compounds. Configuration energies represent a measure of electronegativities. This triangular diagram naturally separates into regions traditionally labeled as metallic, covalent, and ionic [30, 31]. Zintl phases can be assigned to a region between the metallic and ionic corners, but they merge with the covalent region as well. Recently, significant research effort has focused on so-called *polar intermetallics*, which show many features of Zintl phases except that their structures typically do not follow the *8-N rule* [1].

This chapter examines advancing the Zintl–Klemm formalism through theoretical and experimental exploration. As our computational thresholds continue to expand with respect to basis set sizes and improved methodologies, we can investigate

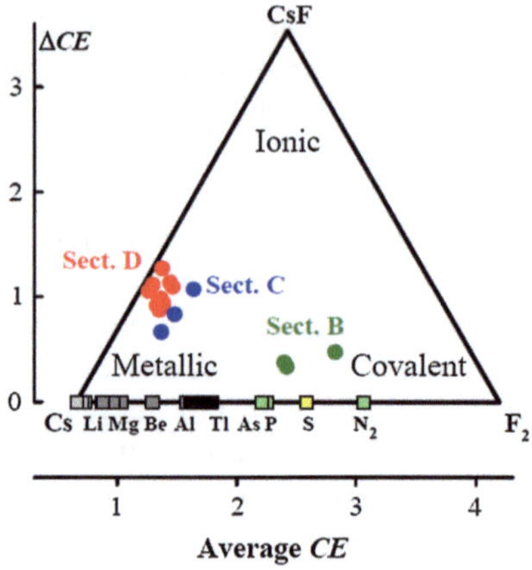

Fig. 1 van Arkel–Ketelaar diagram for binary compounds based on configuration energies (*CE*) [26, 27], which have been renormalized to the Pauling scale of electronegativities. The base of the triangular map consists of all elements ranging from Cs (lowest electronegativity) to F_2 (highest electronegativity); alkali metals, alkaline-earth metals, trielides, pnictogens, and sulfur positions are noted as blocks along this base. The apex is the most "ionic" binary compound, CsF. This map identifies *metallic*, *covalent*, and *ionic* systems. Examples of cases discussed in this review are highlighted in the diagram

increasingly larger molecules and molecular ions with heavier elements, species that are more in line with Zintl ions. In addition, first-principles approaches allow improved quantum mechanical treatment of the active metal components over the semiempirical methods applied in earlier theoretical investigations [24]; hence, we are in a better position to assess the role of cations in the stability and potential reactivity of Zintl phases. We have organized this chapter into sections as follows: Sect. 2 showing the interpretations from ab initio theory on a class of simple molecules demonstrating Klemm's pseudoatom concept – the family of realgars; Sect. 3 exploring the simultaneous role of valence orbitals of the cation and valence electron tuning in a complex system related to MgB_2; Sect. 4 discussing the theoretical assessment of the competition between isotropic (ionic/metallic) forces and anisotropic (covalent) interactions in NaTl and related phases; and Sect. 5 summarizing some future outlooks for the Zintl–Klemm formalism, i.e., why do Zintl phases remain worthy of intense scrutiny and what can be gained from their study?

2 The Pseudoatom Concept: First-Principles Calculations on Realgars

From the theoretical perspective, semiempirical methods suffer because atomic orbital energies are typically input parameters. Although based on atomic spectra, their values can be modified and subsequent analysis of charge partitioning is always controversial. Using ab initio theory would alleviate this concern, but raises other issues, which include the size of the problem that can be studied and the inherent instability of highly negatively charged molecular anions, as often found in Zintl phases. Nonetheless, as quantum mechanical software and computer hardware continue to develop, the application of ab initio theory to Zintl ions, in particular, is on the horizon, and this section demonstrates its utility for a quantitative analysis of energetics and chemical bonding in realgar-type molecular crystals. Although composed on electrically neutral clusters, their structures nicely demonstrate Klemm's pseudoatom concept [21, 22], which can now be studied by the entire range of standard quantum chemical methods, which permit evaluation of energy differences between various structural isomers and full geometry optimization as well as a recently developed bond order analysis to analyze the different structural isomers. It is well known that quantum chemistry has become very powerful for the prediction of structural and energetic information, such as will be presented in this section. However, it remains ever challenging to extract qualitative information about chemical bonding from the resulting wavefunctions, to understand these results. With this challenge in mind, a bonding analysis of such first-principles calculations will also be described qualitatively.

The molecular units in realgar-type crystals are composed of four pnictogen (Pn = N, P, As) and four chalcogen (Ch = S, Se) atoms, i.e., Pn_4Ch_4, arranged in D_{2d} symmetry. There are two distinct atomic positions, one forming a *square planar* arrangement and the other adopting a distorted *tetrahedron* with two atoms above the square plane, and two below it. By considering nearest neighbor contacts, the *square planar* sites are two-bonded whereas the *tetrahedron* sites are three-bonded. According to Klemm's pseudoatom concept [21, 22], the pnictogen atoms should occupy the *tetrahedron* sites and the chalcogen atoms the *square planar* sites. Thus, we adopt a formula convention where the square planar atoms are listed second in the molecular formula so that the predicted formula would be Pn_4Ch_4. As_4S_4 (realgar) and P_4S_4 follow this assignment, but S_4N_4 is observed with the reversed "coloring." An early attempt to use quantum chemistry to understand this "coloring problem" only addresses the three observed structures [32]. Here, we describe a bonding analysis and other quantum mechanical results for all six D_{2d} systems: S_4N_4 vs. "N_4S_4"; P_4S_4 vs. "S_4P_4"; and As_4S_4 vs. "S_4As_4" (in the following, a chemical formula in quotes signifies a hypothetical, unobserved isomer).

Both gas phase [33, 34] and X-ray structures [35–37] are available for tetrasulfur tetranitride, S_4N_4, a shock-sensitive material. Structural parameters depend on the physical state, particularly for the S–S distance, which is 2.666 Å in the gas phase and 2.582 Å in the solid state. The S–N distance is more constant, being 1.623 and 1.616 Å, respectively, for gaseous and solid states. Computation of the S–S distance shows a wide range in results, which depend on the method and basis sets used: density functional theory (DFT) gives, perhaps, the best results to date, using B3LYP/cc-pVTZ to yield 2.751 Å [38]; whereas an earlier theoretical study [39] with the small 6-31+G(d) basis set gave 3.236 Å (MP2) and 2.781 Å (MP4(SDQ)). The bonding in S_4N_4 has been illuminated by force constant analysis of the vibrational data [40–42], as well as evaluated theoretically [32, 39, 43–49], with few cases comparing it to the alternative isomer "N_4S_4" [39, 46]. "N_4S_4" is not observed as an unsubstituted molecule, although early electron diffraction [50] and vibrational experiments [51] mistakenly used this structure in their analysis. A recent theoretical paper [39] has showed that the D_{2d} structure for "N_4S_4" relaxes to C_s symmetry in a ring with seven roughly planar atoms and one N atom puckered out. X-ray structures containing this type of C_s ring structure are found as adducts with Lewis acids, where species such as BF_3 [52], SO_3 [53], or $TeCl_4$ [54] are coordinated to the puckered N atom.

In contrast, tetraphosphorus tetrasulfide, α-P_4S_4 [55, 56], and realgar in two forms, α-As_4S_4 [57] and β-As_4S_4 [58], locate the trivalent pnictogens in the three-bonded, *tetrahedron* positions, and the divalent chalcogens to the two-bonded, *square planar* sites. An early electron diffraction structure of As_4S_4 [33] gives the As–As as 2.49 Å, which is probably spurious, as it is shorter than the 2.59 Å distance in either crystal type, and also is not supported by the theoretical results below. Vibrational data for P_4S_4 and As_4S_4 have also been summarized [59, 60]. Little theoretical work exists for P_4S_4, but there is a recent study of As_4S_4 [61]. No experimental information is available concerning the alternative "S_4P_4" or "S_4As_4" isomers. Note that polycyclic cage isomers, named β-P_4S_4 [55] and pararealgar [62], exist, also containing two *Pn*–*Pn*, eight *Pn*–S, and no S–S bonds, but these are not considered here.

2.1 Methods

To compare the structures and chemical bonding in the two possible isomers, Pn_4Ch_4 or Ch_4Pn_4, closed shell self-consistent field (SCF), i.e., restricted Hartree–Fock (RHF) approaches, second-order perturbation theory (MP2) [63], and DFT using the recently developed, nonempirical meta-GGA functional, known as revised TPSS [64] were used for the geometry optimizations. A large, modern Gaussian basis set is available for all atoms, namely *aug-cc-pVTZ* [65–67], which is specifically designed for the recovery of electron correlation energies.

The chemical bonding analysis is a recently developed, basis set independent method, which will be described more formally elsewhere. It is based on ideas developed for multiconfigurational SCF (MCSCF) calculations, where only up to ca. 16 valence electrons can be studied in detail. The new procedure, described below, can deal with much larger systems and has been implemented in the GAMESS program [68], requiring little more from the user than the input of the molecular geometry.

Quasi-atomic orbitals (QAOs), adapted to the molecular environment, are extracted from the molecular wavefunction, preserving orthogonality, so that each QAO's occupancy and its bonding (population) to neighboring atoms can be easily obtained. The first step toward the orthogonal QAOs is projection of the exact valence (and core) orbitals of the free atoms onto the occupied and virtual orbital spaces of the closed shell SCF wavefunction [69] of the target molecule. The projection of all valence orbitals of the free atom yields, by design, all valence molecular orbitals. After diagonalization of the Fock operator over the initial projected and symmetrically orthogonalized AOs, some eigenvectors are obtained which are *identical* to the occupied SCF valence orbitals, but, in addition, all other valence orbitals which lay in the original canonical orbital virtual space appear [69]. In other words, the unoccupied antibonding valence orbitals are now separated from the remainder of the virtual space of the closed shell SCF calculation. These unoccupied antibonding valence orbitals are what is sought and are termed the valence virtual orbitals (VVOs). The remainder of the virtual space consists of diffuse, polarization, Rydberg, and similar basis functions, which are not of direct relevance to bonding by valence electrons. Localization [70, 71] of both sets of valence orbitals, i.e., the occupied canonical and VVOs, onto the individual atoms gives one orbital per s-block atom, and four per s, p-block atoms. These orbitals must undergo one final step, namely "orientation" along bonding directions. This is accomplished by rotating orbitals on each atom to maximize the bond order of an orbital on one atom to just one orbital on another atom [72]. The resulting oriented, localized, atomic-like orbitals (termed the QAOs) are quite similar to those that would be obtained by similar localization and orientation of a full valence MCSCF calculation, but in very little computer time.

The bond orders and atomic orbital populations are then obtained from a first-order density matrix transformed to this set of oriented, localized, atomic-like orbitals. An ordinary singly and doubly excited configuration interaction (CISD) wavefunction [73], in which the occupied canonical valence MOs are excited into the VVOs gives, at least near a molecule's equilibrium distance, a density matrix which is a good approximation to that of a full CI result in the valence orbital space. This density permits us to understand the difference between full bonding (occupations of 2 and 0 in bonding and antibonding orbitals) and partial bonding, where the antibonding orbitals are partially occupied. Transforming the CISD density matrix to the carefully prepared oriented atomic-like orthogonal orbitals permits Coulson's bond-and-charge order analysis [74, 75]. Since the oriented, atomic-like orbitals are orthonormal, the diagonal elements of this oriented, localized, orbital density matrix immediately have the property of summing to the total number of valence electrons, and each is therefore an electron population for that particular molecule-adapted atomic orbital. Summing all orbitals on a single atom thus gives directly an atomic population analysis (charge). The off-diagonal elements of the density have the properties of a bond order: they are bounded between $+1$ (fully bonding) and -1 (fully antibonding) [72], so bond orders may simply be read out from the first-order density matrix. Hence, the designation, *charge and bond order matrix* [76], common prior to 1955 [77], is entirely appropriate to the density matrix expressed over these oriented, localized, atomic-like orbitals.

It is worth noting that the VVOs are themselves useful, as they are a quantitative embodiment of the LUMO concept. The normal canonical orbitals are not convergent with respect to the basis set. In fact, for neutral molecules the normal virtual space spans a continuum of free electron states. This procedure is also basis set independent, in which the projected orbitals are designed to be valence in nature, and thus converge to a limit as the AO basis is taken to the extreme.

2.2 Results

Table 1 contains structure and energy data for Pn_4Ch_4 and Ch_4Pn_4 systems, both in D_{2d} symmetry. The signs for each energy difference indicate that the three levels of theory displayed easily predict the observed structural isomer. In fact, none of the unobserved molecules is stable in D_{2d} symmetry, but, rather, each distorts to C_s symmetry (shown in Fig. 2), which were found to be minima by hessian

Table 1 Comparison of computed and experimental structures, with computed energy differences for realgar-type Pn_4Ch_4 and Ch_4Pn_4 molecules

	—S_4N_4—		—"S_4P_4"—		—"S_4As_4"—	
	d(S–S)	d(S–N)	d(S–S)	d(S–P)	d(S–S)	d(S–As)
RHF	2.446	1.594	3.042	2.074	3.304	2.197
DFT	2.747	1.630	3.198	2.099	3.314	2.217
MP2	3.042	1.634	3.399	2.101	3.486	2.234
ED	2.666	1.623				
XRD	2.58	1.61				
	—"N_4S_4"—		—P_4S_4—		—As_4S_4—	
	d(N–N)	d(N–S)	d(P–P)	d(P–S)	d(As–As)	d(As–S)
RHF	1.551	1.731	2.329	2.112	2.555	2.228
DFT	2.948	1.660	2.392	2.213	2.626	2.244
MP2	3.301	1.670	2.360	2.105	2.599	2.227
ED					2.49	2.23
XRD (α)			2.360	2.118	2.57	2.24
XRD (β)			–	–	2.593	2.243
ΔE (RHF)	+58.2		−164.9		−159.9	
ΔE (DFT)	+54.6		−81.3		−81.0	
ΔE (MP2)	+34.1		−59.4		−57.4	
C_s: ΔE (MP2)	+22.5		−53.4		−49.3	

Distances are in Å; energies are in kcal/mol, $\Delta E = E(Ch_4Pn_4) - E(Pn_4Ch_4)$, both in D_{2d} symmetry. See text for the meaning of the final row. *ED* electron diffraction, *XRD* X-ray diffraction

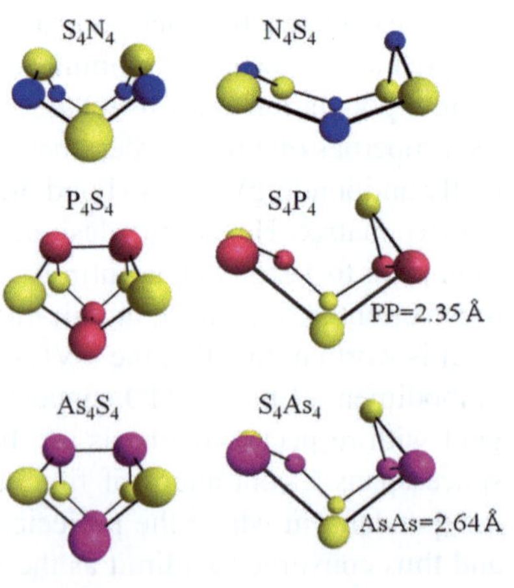

Fig. 2 MP2/aug-cc-pVTZ geometries of observed molecules on the left, and unobserved forms after relaxation from D_{2d} into C_s on the right. N_4S_4 closely resembles known adducts [52–54], and the previous work of Chung and Lee [39]. No other conformations or puckerings were explored for S_4P_4 or S_4As_4, which may not be at their global minima

computation at the MP2 level. The energy differences between each pair of structural isomers become smaller after relaxation of the unobserved D_{2d} isomer to C_s symmetry is allowed (final row of Table 1). With respect to predictions of interatomic distances, MP2 is slightly better than DFT, both of which produce values close to experiment for P_4S_4 and As_4S_4. As other workers have found [39], however, DFT is better than MP2 for the structure of S_4N_4 (the fairest comparison of theory is to the gas phase electron diffraction structure), but neither is very good. Accordingly, MP2 geometries in D_{2d} symmetry will be used for the bonding analysis and comparisons of all six species, but in addition, the DFT and experimental electron diffraction geometry will be used for S_4N_4.

The bonding analysis involves constructing four oriented orthogonal orbitals on each atom, whose corresponding density matrix will contain charges and bond orders. Figure 3 shows some of these orbitals. At the MP2 geometries, the 32 valence orbitals group into six symmetry equivalent types. Each tetrahedral atom (symbolized *T*) has one lone pair, one sp^3-like hybrid directed at another tetrahedral atom, and two equivalent sp^3-like hybrids directed toward a square planar atom (symbolized *S*). The latter two orbitals always participate in strong bonding, while the orbital directed toward a tetrahedral atom varies in its bond participation. Each of the square planar (*S*) atoms has two lone pair orbitals, which are not symmetry equivalent, and two sp^3-like hybrids directed toward tetrahedral atoms, which are symmetry equivalent. Figure 3 shows these orbitals for S_4N_4. Their shapes in the other five species are very similar. Table 2 contains data supporting a Mulliken population analysis [32] on the three observed molecules (S_4N_4, P_4S_4, As_4S_4), a study which concluded that "the tetrahedral atoms are at the positive ends of the bonds, and the square is a net electron acceptor in each case, the amount varying with the electronegativity difference." But these results afford an even deeper insight into the bonding of all six species.

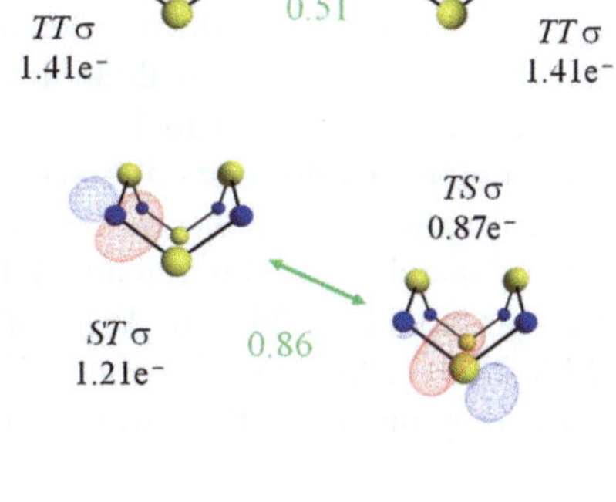

Fig. 3 Quasi-Atomic Orbitals (QAOs) in S_4N_4, with their electron populations and bond orders listed for the MP2 geometry. The two lone pairs at each *S* site (not shown) contain $1.72e^-$. The S–S bond order is distance sensitive (see text). The total valence electron count is $44e^-$ ($= 4 \cdot 1.99e^- + 4 \cdot 1.41e^- + 8 \cdot 1.21e^- + 8 \cdot 0.87e^- + 4 \cdot 1.72e^- + 4 \cdot 1.72e^-$)

Table 2 Electron populations by orbital and by atom, and bond orders in T_4S_4 realgar-type clusters

	Orbital occupancies and atomic charges								Bond orders		
	Tetrahedral atom (*T*)				Square planar atom (*S*)						
T_4S_4	*TT* σ	*TS* σ	*lp*	*Q*(*T*)	*ST* σ	*lp*′	*lp*″	*Q*(*S*)	*TT*	*TS*	Seen?
"N_4S_4"	1.35	1.21	1.93	+0.30	0.87	1.97	1.59	−0.30	0.41	0.88	N
S_4N_4	1.41	0.87	1.99	+0.86	1.21	1.72	1.72	−0.86	0.51	0.86	Y
"S_4P_4"	1.65	1.29	1.98	−0.21	0.76	1.95	1.32	+0.21	0.23	0.86	N
P_4S_4	1.07	0.72	1.97	+0.52	1.33	1.99	1.87	−0.52	0.86	0.87	Y
"S_4As_4"	1.69	1.31	1.98	−0.29	0.73	1.97	1.28	+0.29	0.18	0.86	N
As_4S_4	1.07	0.71	1.98	+0.53	1.34	1.98	1.87	−0.53	0.86	0.86	Y

The symbols *T* and *S* refer, respectively, to tetrahedral and square planar sites. Except in S_4N_4, *lp*′ refers to a lone pair with *s*/*p* hybridization, while *lp*″ is nearly *p* in character

The present bonding analysis assigns an electron occupation number to each orbital, for example, in S_4N_4, the tetrahedral sulfur's lone pair contains 1.99e$^-$. Summing the population of all four orbitals on this sulfur gives 5.14e$^-$ [= 1.41e$^-$ + 2(0.87e$^-$) + 1.99e$^-$], so its total charge *Q*(S) is +0.86e$^-$. Of course, the charge on the square planar nitrogen atoms is the exact opposite of the tetrahedral atom's charge. Therefore, S_4N_4 is most of the way to $(S^+)_4(N^-)_4$! Indeed, the molecular form that is experimentally realized in all cases is the one where the *S* atom is the more negatively charged than the *T* atom [32]. For the two cases, Pn_4S_4 and S_4Pn_4, which have the greatest energy difference (*Pn* = P, As), the *S* site is actually positive in the unobserved forms. Even for *Pn* = N, the *S* site is >1/2 charge unit more positive in the higher energy form.

The lone pair orbital at the *T* site is always a full electron pair, in keeping with *T* being a trivalent site. The two orbitals at the *S* site, which are labeled *lp*′ and *lp*″ in Table 2, consist of, respectively, an sp^2-hybrid and a *p*-type orbital. S_4N_4 is the only exception, since the N atom has two sp^3 lone pairs with equal occupancy. The experimentally realized structures are those where the *S* site's lone pairs total approximately 4e$^-$, meaning *S* is a divalent site. Accordingly, P_4S_4 and As_4S_4 are cases where the pnictogen and chalcogen adopt their normal chemical valencies, while S_4N_4 can be considered as trivalent S^+ and divalent N^- (the actual charges are ±0.86e$^-$, with 3.44 lone pair electrons at N). The large atomic charge assignments for S_4N_4 confirm the S^+/N^- structures sometimes drawn by some other workers [36, 43–45].

Consideration of the bonding orbitals indicates that the *S* site atom uses a sp^2-hybrid, called "*ST* σ," which is populated by 0.73–1.34e$^-$ in the six species, to bond at each of two *T* atoms with another sp^2-hybrid, denoted "*TS* σ," whose electron count ranges from 0.71 to 1.31e$^-$ in the six species. In all six molecules, the *ST* bond order (penultimate column) is essentially the same, ≈0.87, which is a fairly typical value for a polarized single bond. The polarity is reversed in direction when *Pn* and *Ch* are interchanged, but its magnitude is not: compare 1.31/0.73e$^-$ in S_4As_4 to 0.71/1.34e$^-$ in As_4S_4. The sum of the "*ST* σ" and "*TS* σ" columns is 2.03–2.09e$^-$. In short, the eight *ST* bonds are entirely normal, in all six species, formed by one pair of electrons polarized toward the more electronegative atom.

The polarization, i.e., partial ionicity, decreases from the maximum bond order of 1 for a homopolar covalent situation to about 0.87 for each, as does the slight excess over 1 pair, which resides in the antibonding combination of these two QAOs.

In contrast, the two *TT* bonds in each molecule are much more interesting. In the three observed species, each "*TT* σ" has close to one electron: $T = S$, 1.41e$^-$; $T = P$, 1.07e$^-$, $T =$ As, 1.07e$^-$, and can therefore bond well, with *TT* bond orders of SS = 0.51, PP = 0.86, and AsAs = 0.86. The latter two values are respectable fractions of a single bond and are equal to the *ST* bond orders. However, the most interesting case of the SS bond in S_4N_4 is noticeably smaller (0.51), due to its atomic "*TT* σ" contributing more than one electron (1.41). Nitrogen's inability to hold two full electron pairs, noted above, is the source of the sulfur's "*TT* σ" excess electron count, causing sulfur to be somewhat less than trivalent. In the three unobserved species, each "*TT* σ" contributes well over one e$^-$: N = 1.35e$^-$, S = 1.65e$^-$, S = 1.69e$^-$, so the *TT* bond order is small: 0.41 in N_4S_4, 0.23 in S_4P_4, and 0.18 in S_4As_4. Of course, these much smaller bond orders are due to extensive contributions of *TT* antibonding orbitals to contain the total e$^-$ count in excess of two.

An interpretation of the chemical bonding in the S/P and S/As isomer pairs is clear-cut. When the more electronegative sulfur atom takes the *T* site, its sp^2-hybrid directed toward the other sulfur is forced by nature to have about 1.67e$^-$, and thus a very low *TT* bond order of about 0.20. Not surprisingly, these are strongly disfavored energetically. The S/N system is more ambiguous because sulfur is not the electronegative atom. By approaching S^+/N^- charges, sulfur's "*TT* σ" with 1.41e$^-$ and a *TT* bond order of 0.51 in S_4N_4 has only slightly more *TT* bonding than N_4S_4 nitrogen "*TT* σ" =1.35e$^-$ and *TT* bond order of 0.41. The S/N system shows the smallest energy difference, with the observed structure (S_4N_4) being the case where the *TT* bond order is largest.

The sulfur–sulfur bond in S_4N_4 is the weakest *T*–*T* bond, by all measures. This includes experiment, where the S–S distance is found to be 0.09 Å longer in the gas vs. solid phases, a sure sign of a low force constant. DFT and MP2 calculations predict S–S distances 0.08 and 0.38 Å shorter than the measured gas phase value, respectively (To date, no calculation has correctly predicted the S–S distance, indicating CCSD(T) should be tried). Results in Table 3 were obtained at the MP2 geometry, with a rather long S–S distance. The same bond order analysis may be attempted at the DFT and experimental geometries to explore the dependence of the S–S bonding on distance. Unfortunately, the localization of the orbitals onto atoms at either DFT or experimental geometries leads to loss of exact symmetry equivalence in the localized sets, producing two distinct pairs of *Pn* atoms and two pairs of S atoms. This artifact is more pronounced at the shortest S–S distance (experiment), where orbital populations split into *TT* σ = 1.30 or 1.37e$^-$, *TS* σ = 0.88 or 0.90e$^-$, and *ST* σ = 1.18 or 1.20e$^-$. The SS bond order between *TT* σ orbitals with 1.30 and 1.37e$^-$ increases to 0.58 from its value 0.51 at the MP2 geometry. The intermediate DFT geometry has an intermediate SS bond order, 0.56. The S–S bond is clearly weaker than the 0.86 value found for the *PnPn* bond order in P_4S_4 and As_4S_4, even at the observed S–S distance.

Many theoretical and experimental bonding analyses have been performed on S_4N_4. Analyses of vibrational spectra have assigned the S–S force constant in S_4N_4 as 2.1 mdyn/Å [42], much smaller than S–N's 33.2 mdyn/Å. The Atoms in Molecules scheme [44, 47] and the related electron localization function (ELF) [49] find bond critical points within the S–S contact, but with smaller magnitudes of electron density than for the critical points of the S–N bonds. In fact, this had already been observed [36] experimentally in X-ray data! The Mayer bond order, which unlike the present method is basis set dependent, was found to range from 0.22 to 0.50 for SS, and 1.09 to 1.35 for SN [48]. The large S^+/N^- charge assignment in the present work is also in good agreement with previous Mulliken population analyses [32, 43], a procedure that is also basis set dependent. There is general agreement, therefore, between the present work and previous experimental and theoretical papers, to the effect that the S–S bond is a rather weak single bond formed between two approximately trivalent S^+ ions. The low S–S bond order is certainly related to the shock sensitivity of S_4N_4.

Other workers have discussed S_4N_4 in terms of *d*-orbital participation, delocalized bonding, or partial π-bonding. For example, Schugart lists the following resonance structures as contributors [46]:

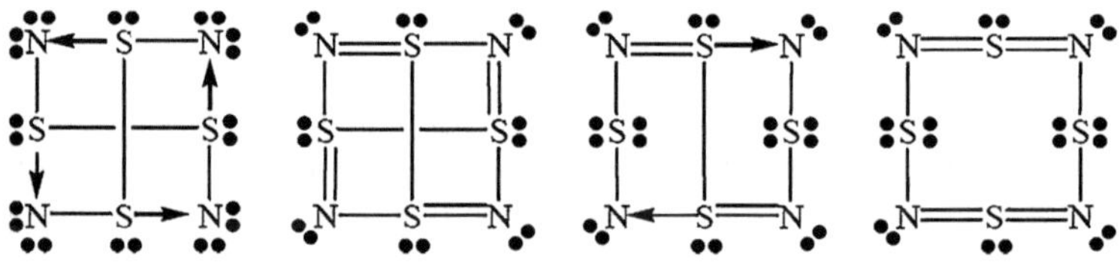

The first is typically drawn with S^+ and N^- atoms, rather than Schugart's dative bonds, but is the structure originally proposed by Lindqvist in 1958 [82], as well as the model supported by the present bonding analysis. What about the other resonance diagrams? The procedure used here yields small bond orders, particularly between the N lone pairs and the S atom's bonding orbitals, namely −0.33, 0.21, 0.10, 0.10, −0.31, and 0.10 which sum to a small net antibonding: −0.13. These interactions are also present in P_4S_4 and As_4S_4, but are halved in size. Thus, the net interaction between the *T* and *S* sites consists of a two-electron σ-bond. There are only negligible interactions between two *S* site atoms, or between the two *T* site atoms that are not directly bonded.

The observed species, S_4N_4, P_4S_4, and As_4S_4, are found to contain ordinary two-center, two-electron bonds between trivalent *T* sites and divalent *S* sites, which have, respectively, one and two lone pairs. Analysis of the chemical bonding using a procedure developed for ab initio calculations, a procedure which is also independent of orbital basis set, shows that all three species obey the Zintl–Klemm formalism. P_4S_4 and As_4S_4 segregate the pnictogen atoms onto the trivalent *T* sites and S atoms on the divalent *S* sites. S_4N_4 also fulfills this formalism by approximating to nearly full charge separation $(S^+)_4(N^-)_4$, with 3.44 rather than 4 lone pair electrons at N. Its S–S bond is weak, since sulfur's charge of +0.86 precludes full trivalency. Although the "coloring" of the realgar structure may, at first glance, seem counterintuitive, indeed, the observed structure is in line with the elemental electronegativities.

3 Role of the Cation and Valence Electron Tuning: MgB_2 Types

Numerous ternary, quaternary, and other multinary Zintl phases have been targeted and successfully prepared from (semi)metallic elements of different electronegativities [24]. Most multinary cases involve a single active metal with combinations of electronegative metals, e.g., Na–Zn–Sn [83, 84], K/Rb–In–Pb [85], Na–Tl–Sb [86], K–Pb–Bi [87], Sr–Al–Ge [88], *Ae*–Al–Sn (*Ae* = alkaline earth metal) [89], Ca–In–Sn [90], and *AE*–Sn–Bi [91]. Fewer examples arise from combinations of active metals with a single electronegative element, viz., Ca–Ba–Sn [92], Ca/Sr–Ba–Pb [93], and Sr–Ba–Bi [94], or combinations of both species, e.g., Ca–Eu–In–Ge [95], Sr–Ba–Cd–Sb [96], and K–*AE*–Cd–Sb [97]. Astonishingly, for many of the former cases, the binary phase diagrams of several pairs of electronegative (semi)metals, e.g., Zn–Sn, Al–Ge, Al–Sn, fail to exhibit any compounds in the solid state [98], but in the presence of an active metal yield numerous phases with potentially interesting physical properties. As the combinations of electronegative elements approach the Zintl border, i.e., the line separating triel elements (B, Al, Ga, In, Tl) from tetrel elements (C, Si, Ge, Sn, Pb) [24, 25], the structures begin to deviate from the Zintl–Klemm formalism. The two primary *aristotypic* structures upon which recent applications of the Zintl–Klemm formalism are based are the two-dimensional (2D) graphite structure and the three-dimensional (3D) diamond structure. Since recent efforts have emphasized the role of the active metal, which is formally the "cation" in these phases, the next two sections will examine the impact of the active metal, with respect to valence atomic orbitals and size, on these two aristotypes, i.e., substances whose structures are based on MgB_2 and NaTl.

Species related to magnesium diboride engendered renewed enthusiasm and interest during the past decade due to the discovery of an unusually high superconducting transition temperature at 39 K in MgB_2 itself [99]. Applying the Zintl–Klemm formalism to MgB_2 rationalizes the 2D graphite-type (6^3) sheets of boron, $[B_2]^{2-}$, which are isoelectronic to carbon. These planes stack in an eclipsed fashion, unlike various polymorphs of graphite, 2H [100–102] and 3R [103], with Mg atoms located in all hexagonal prisms between the boron sheets. The key feature leading to the superconducting properties of MgB_2 involves strong coupling between partially occupied B–B σ-bonding states and in-plane vibrational modes of boron [104]. Attempts to alter the position of the Fermi level by doping the boron sites with carbon atoms or the magnesium sites with aluminum or lithium atoms invariably lower the transition temperature [105–112]. On the other hand, superconductivity has been observed in CaAlSi [113], which is a ternary variant of the MgB_2 type, but this example has nine valence electrons, which would make it isoelectronic with nonsuperconducting AlB_2. A thorough systematic investigation of *AeTrTt* systems, *Ae* = alkaline earth metal, *Tr* = triel element (Al, Ga, In), and

Tt = tetrel element (Si, Ge, Sn), revealed superconducting behavior for many nine-electron ($9e^-$) systems [114–120]. Theoretical investigations identified partially occupied *Tr–Tt* π^* and *Ae d* states in conjunction with a soft phonon mode that is associated with out-of-plane *Ae–Tt* vibrations as the origin for the superconducting behavior [121–125]. With respect to the Zintl–Klemm formalism, a $9e^-$ graphite sheet is electron-rich, a situation that could lead to structural distortion. Of course, the "cations" create both electronic and elastic forces that can counteract the driving forces for distortion of the graphite-like net. Thus, $A(X_{1-x}X'_x)_2$ systems create a fertile garden of compounds to examine the interrelationships among atomic sizes, valence electron concentrations, and the role of cations *A* on structure–property relationships in Zintl phases within the Zintl–Klemm concept.

There are numerous structural variations observed for ternary $A(X_{1-x}X'_x)_2$ phases derived from the AlB_2-type structure, alternatives which involve puckering of the hexagonal nets, interconnections between these nets, and colorings of the electronegative metals within each hexagonal net, some of which are illustrated in Fig. 4 [126]. In general, planar sheets occur for $8e^-$ systems; puckered sheets with interconnections are found for systems less than eight-electrons; and puckering without interconnections occur for systems above eight-electrons. Among $8e^-$ systems, in particular, puckering leads to 3D diamond, sphalerite, or wurtzite-type networks when the scaled *c*/*a* ratio,[1] labeled as $(c/a)^*$, drops below a threshold value of 0.85; planar 2D nets remain when this *c*/*a* ratio remains above this threshold [127]. But, as mentioned above, $9e^-$ systems are particularly interesting, especially considering the interaction between the cation and 2D sheets. In the remainder of this section, we will focus on $Eu(X_{1-x}X'_x)_2$ phases, where *X* and *X′* are restricted to Zn, Ga, or Ge, which are selected to emphasize the electronic effects associated with the 2D components while minimizing atomic size differences. In all phases, Eu exhibits divalent behavior according to numerous physical measurements, including magnetic susceptibility and X-ray near edge spectroscopy. At certain points, we will discuss the influence of atomic sizes in the 6^3 nets by considering Si or Sn in place of Ge, as well as examine the role played by valence orbitals at Eu.

Common AX_2 structures for *A* coming from the electropositive divalent elements (Ca, Sr, Ba, Eu) and *X* coming from the tetrelides (Group 14), trielides (Group 13), and dodecalides (Group 12) include the orthorhombic KHg_2 type (also called the $CeCu_2$ type), the hexagonal $CaIn_2$ type, the hexagonal AlB_2 type, as well as the three Laves phase structure types: cubic $MgCu_2$ type and the hexagonal $MgZn_2$ and $MgNi_2$ types [128]. A recent analysis of $6e^-$ and $8e^-$ AX_2 phases indicates that AlB_2-type structures are preferred for large atomic size ratios (V_A/V_X) and electronegativity differences ($\chi_X - \chi_A$), whereas the KHg_2-type structure becomes preferred as the corresponding electronegativity difference becomes smaller while the

[1] In these AlB_2-related $A(X_{1-x}X'_x)_2$ structures, $(c/a)^*$ uses c = average near neighbor $A\cdots A$ distance along the stacking axis and a = average near neighbor $A\cdots A$ distance orthogonal to the stacking axis.

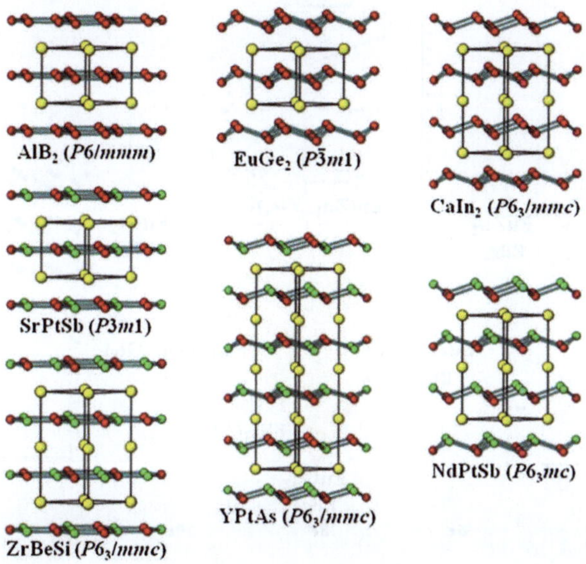

Fig. 4 Various binary and ternary structure types based on planar or puckered 6^3 nets of electronegative (semi)metals derived from the AlB_2-type. SrPtSb- and ZrBeSi-types involve alternant 6^3 nets and vary in how adjacent layers are stacked along the *c*-axis. $CaIn_2$- and NdPtSb-types show puckered 6^3 nets, a distortion which creates short interlayer distances and local coordination that is distorted tetrahedral. YPtAs-type is a 1:1 ternary intergrowth of NdPtSb- and $EuGe_2$-types

size ratio remains constant [129]. Laves phases dominate when these two parameters are smaller, which has been interpreted as effective packing of differently sized metal atoms [130–132]. The 12-coordinate metallic radius of Eu is 2.04 Å, which lies between that of Ca (1.97 Å) and Sr (2.15 Å) [133], and Eu is slightly more electronegative than the *Ae* elements according to both Pauling's scale and Allen's configuration energies [27] (Eu: ca. 1.10; Ca: 1.03; Sr: 0.96; Ba: 0.88 in Pauling units). For the $Eu(X_{1-x}X'_x)_2$ series ($X, X' =$ Zn, Ga, Ge), there is a large size ratio (V_{Eu}/V_X) and an increasing electronegativity difference ($\chi_X - \chi_{Eu}$) from Zn to Ge, which implies a switch from orthorhombic KHg_2-type to hexagonal AlB_2-type phases as the valence electron count varies from 6e$^-$ ($EuZn_2$) to 8e$^-$ (EuZnGe). Figure 5 summarizes recent results on $Eu(X_{1-x}X'_x)_2$ phases: investigations have extensively studied $Eu(Zn_{1-x}Ge_x)_2$ and $Eu(Ga_{1-x}Ge_x)_2$ phases, but few investigations of $Eu(Zn_{1-x}Ga_x)_2$ exist.

3.1 The $Eu(Zn_{1-x}Ga_x)_2$ System

Other than the binary end-members of this series, i.e., 6e$^-$ $EuZn_2$ [134] and 8e$^-$ $EuGa_2$ [135], only 7e$^-$ EuZnGa is reported [136], each of which adopts the

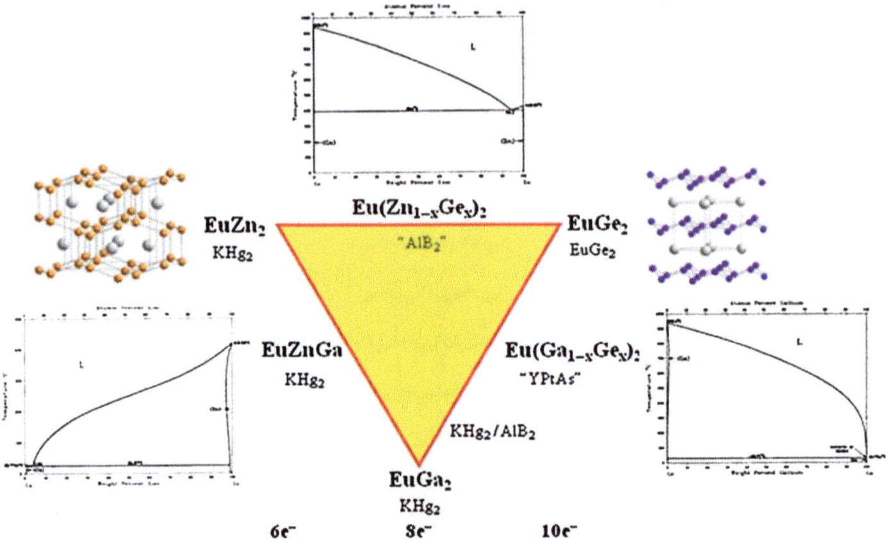

Fig. 5 Summary of phase behavior in the Eu($X_{1\text{-}x}X'_x$)$_2$ systems, X and X' = Zn, Ga, or Ge. The structure types for each binary, $EuZn_2$, $EuGa_2$, and $EuGe_2$ are illustrated, as are the binary Zn–Ge, Zn–Ga, and Ga–Ge phase diagrams, which indicate no compound formation. The text will address each side of the triangular diagram in more detail

orthorhombic KHg_2-type structure, illustrated in Fig. 5. As the valence electron count increases, the unit cell volume decreases. The electronegative metals form a four-bonded–three-dimensional (4b–3D) net based on eclipsed stacking of puckered pseudo-hexagonal 2D layers, a stacking which creates zigzag ladders of four-membered rings. The four-bonded coordination at Zn or Ga atoms is a distorted tetrahedron with three distinct distances, with the longest bond directed between the 2D layers, and two bond angles fixed at 90°. An earlier theoretical assessment of this "versatile" structure type [137] indicated its increasing stability from 4e$^-$ to 7.5e$^-$ per formula unit [138]. Figure 6 displays the electronic density of states (DOS) curve and nearest neighbor T–T and Eu–T crystal orbital Hamilton population (COHP) [139] curves for T = Zn calculated using a tight-binding, linear muffin–tin orbital (TB-LMTO) code [140]; the Fermi level (E_F) for 6e$^-$ $EuZn_2$ is the reference energy value. E_F falls in a narrow "valley" of the total DOS curve, called a "pseudogap", a feature that lies ca. 1.6 eV below a deeper pseudogap corresponding to 8.0 valence electrons. According to the T–T COHP curves, this deep pseudogap occurs at the crossover between T–T bonding states below and T–T antibonding states above. Thus, 8e$^-$ $EuGa_2$ would satisfy the Zintl–Klemm formalism, although the presence of four-membered rings in its structure violates concepts from the method of moments [141], which would predict an alternative structure with just six-membered rings such as the $CaIn_2$ type. In this regard, the "cations" play a significant role. Both the DOS and Eu–T COHP curves reveal significant Eu–T bonding between 6.0 and 8.0 valence electrons; Eu valence

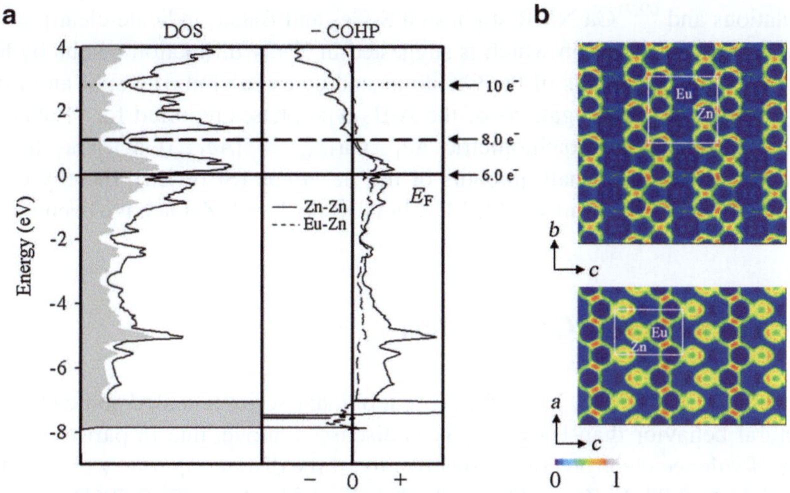

Fig. 6 (**a**) DOS and COHP curves for $EuZn_2$. (Left) total DOS (solid line), Eu PDOS (white region), and Zn PDOS (gray region). (Right) Eu–Zn and Zn–Zn COHP curves. E_F for $6e^-$ $EuZn_2$ is indicated by the solid line. The numbers of valence electrons that correspond to various optima of the Eu–Zn and Zn–Zn COHP curves are noted. (**b**) ELF distribution depicted as filled contour slices, for $EuZn_2$. Left: (1 0 0) slice, $x = 0$. Right: (0 1 0) slice, $y = 0.06$

orbitals, primarily 5*d* and 6*s*, contribute ca. 50% to the total DOS in this region, while Eu constitutes just 33 atomic percent in EuT_2. For the case of $6e^-$ $EuZn_2$, Eu states contribute more than 33% to the total DOS between −2.8 eV and E_F. Another measure of the influence of the Eu atoms is seen for the Zn 3*d* orbitals: these occur as a narrow (0.6 eV wide) band centered at ca. 7.50 eV below E_F, and, thus, can be treated as formally filled. In the theoretical calculation, when the Zn 3*d* orbitals are specifically treated as filled, noninteracting *core* orbitals, their energy value is 7.13 eV below E_F. Thus, a Brewer-type empty *d*-filled *d* orbital interaction, viz. Eu 5*d*–Zn 3*d* [142], also play a stabilizing role for $EuZn_2$ by ca. 7.3 eV (ca. 170 kcal/mol).

The $6e^-$ $AeZn_2$ (*Ae* = Ca, Sr, Ba) are isostructural with $EuZn_2$ under ambient conditions, and transform to the hexagonal Laves phase $MgZn_2$ type under elevated pressure [132]. On the other hand, the $8e^-$ $AeGa_2$ adopt different structures than $EuGa_2$; $CaGa_2$ forms the hexagonal $CaIn_2$ type, whereas X-ray diffraction studies of $SrGa_2$ and $BaGa_2$ suggest the hexagonal AlB_2 type [143]. Analysis of the ELF in $CaGa_2$ produced two attractors, each located at the centers of the inequivalent Ga–Ga contacts: the corresponding integrated electron densities vary inversely with the interatomic distances. In $SrGa_2$ and $BaGa_2$, the ELF analysis revealed different attractor topologies than in $CaGa_2$: two maxima occur above and below each Ga–Ga contacts within the planar 6^3 nets, and additional attractors are found directly above and below each Ga atom. These attractors are attributed to lone pair behavior, but they also occur near the centers of distorted tetrahedra formed by Ga and three *Ae* atoms, indicative of multicenter bonding. Further electronic structure calculations and ^{69,71}Ga NMR studies on $SrGa_2$ and $BaGa_2$ indicate clear puckering of the 6^3 nets, a distortion which is suggested in X-ray diffraction results by highly anisotropic displacements of the Ga atoms and consideration of a split-atom model [143]. Additional investigations of the AlB_2-type phases reported for "$EuGa_2$" and "$SrGa_2$" revealed nonstoichiometric $Ae_{1-x}Ga_{2+3x} \triangleq [Ae_{1-x}(Ga_3)_x]Ga_2$ in which Ga_3 triangles replace small amounts of the *Ae* atoms (Sr or Eu), thereby creating some four-bonded Ga atoms [144, 145]. Finally, no $7e^-$ *Ae*ZnGa have been reported.

3.2 The $Eu(Zn_{1-x}Ge_x)_2$ System

A systematic study [146, 147] of this pseudobinary series revealed somewhat richer structural behavior than $Eu(Zn_{1-x}Ga_x)_2$ discussed above, due in part to the wider range of valence electron count. Specifically, three distinct regions were identified: (1) x close to 0.00, $EuZn_{2-\delta}Ge_\delta$, orthorhombic KHg_2 type; (2) $0.50(1) \leq x \leq 0.68$ (1), hexagonal AlB_2 type, which we can formulate as $EuZn_{1-y}Ge_{1+y}$ $(0 \leq y \leq 0.36)$; and (3) x close to 1.00, $EuZn_\delta Ge_{2-\delta}$, trigonal $EuGe_2$ type. Regions not covered by these ranges showed two-phase behavior, except for the special situation that emerged at $x = 0.75$, i.e., $9e^-$ $EuZn_{0.5}Ge_{1.5}$. A striking structural consideration of the $Eu(Zn_{1-x}Ge_x)_2$ series is the monotonic increase of the $(c/a)^*$ ratios and unit cell volumes as Ge content increases for the three different regions, while the 6^3 nets vary from puckered in $EuZn_2$ to planar in $EuZn_{1-y}Ge_{1+y}$, and then to puckered in $EuGe_2$ (Fig. 7). The trend in unit cell volumes counters expectations based on the relative sizes of Zn and Ge, whether 12-coordinate metallic or covalent radii are used. However, the increasing $(c/a)^*$ ratios suggest an increasing repulsion between adjacent $^2_\infty[(Zn_{1-x}Ge_x)]$ 6^3 nets. The relationship between the puckering of the 6^3 nets and the $(c/a)^*$ ratios also depends on the electronic structure [127]. At low $(c/a)^*$ ratios, puckering is driven by an increasing occupation of intraplanar π-antibonding orbitals through interplanar p–p σ overlap. As the $(c/a)^*$ ratio increases, this interplanar σ overlap steadily decreases and the 6^3 network becomes planar, as in graphite. The puckering observed in $Eu(Zn_{0.25}Ge_{0.75(2)})_2$ and $EuGe_2$, however, can be attributed to the local electronic structure at Ge, which can be rationalized by the Zintl–Klemm concept.

On the Zn-rich side, X-ray powder diffraction data suggested that Ge atoms could replace small amounts of Zn atoms in orthorhombic $EuZn_2$ [146]. Although unit cell volumes remain essentially constant with increasing Ge content (blue points in Fig. 7a), the small variations in a and c axis lengths and phase analysis indicate an upper bound on Ge substitution to be $x < 0.2$. According to the DOS curve in Fig. 6 for $EuZn_2$, $EuZn_{2-\delta}Ge_\delta$ continues to occupy bonding states, but the DOS values at the new Fermi levels increase monotonically, a feature which is indicative of greater electronic instability. However, the system opts to form two distinct phases rather than exhibit any structural or electronic distortion. Calculations on model KHg_2-type structures for the composition "$EuZn_{1.5}Ge_{0.5}$" indicated that the maximum Ge substitution in $EuZn_{2-\delta}Ge_\delta$ lies between $\delta = 0.24$ $(x = 0.12;$

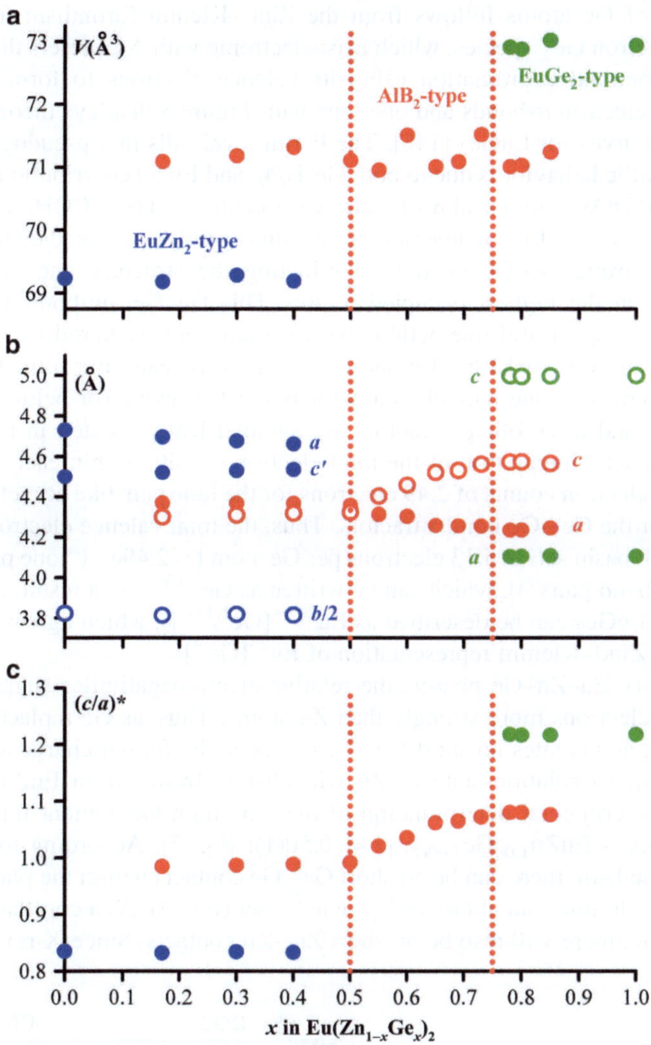

Fig. 7 Variations in structural parameters for $Eu(Zn_{1\text{-}x}Ge_x)_2$ vs. loaded Ge content, x, based on results from X-ray powder diffraction. (**a**) Volumes per formula unit; (**b**) Lattice constants; and (**c**) $(c/a)^*$ ratios. Black/blue symbols: orthorhombic $EuZn_2$-type phases; light gray/red symbols: hexagonal AlB_2-type phases; and dark gray/dark green symbols: trigonal $EuGe_2$-type phases. The single phase region, $EuZn_{1-x}Ge_{1+x}$, occurs between the dotted light gray/red lines on these graphs

$6.48e^-$) and $\delta = 0.62$ ($x = 0.31$; $7.48e^-$), a conclusion based on Zn–Ge and Zn–Zn COHP analysis [147]. Furthermore, the lowest total energy is obtained when there are no Ge–Ge near neighbor contacts, such that Ge behaves formally as "Ge^{4-}."

At the Ge-rich end of the $Eu(Zn_{1-x}Ge_x)_2$ series, there is no evidence to suggest replacement of Ge by Zn, i.e., $EuZn_\delta\,Ge_{2-\delta}$. Nevertheless, in $EuGe_2$, the puckered

3b–2D net of Ge atoms follows from the Zintl–Klemm formalism for $[Ge_2]^{2-}$. The five-electron Ge^- species, which is isoelectronic with As, prefers three-bonded trigonal pyramidal coordination using its valence electrons to form three two-center, two-electron σ-bonds and one lone pair. Figure 8 displays theoretical DOS and COHP curves for $EuGe_2$ [148]. The Fermi level falls in a pseudogap, and the weakly metallic behavior is due to both Ge 4*s*/4*p* and Eu 5*d* contributions; note that the occupied DOS region is almost entirely Ge-centered. The −COHP curve shows that both Ge–Ge and Eu–Ge interactions are almost fully optimized up to the Fermi level, with strong Ge–Ge bonding dominating the structure and only slightly antibonding in the highest occupied region. This Ge–Ge antibonding character arises from *sp–sp* orbital interactions between adjacent puckered Ge sheets. The topological analysis of ELF, also shown in Fig. 8, reveals four attractors around each Ge atom: one lone pair-like attractor is located above (or below) Ge atoms along [001] and three other attractors are symmetrically located in the midst of Ge−Ge contacts. Integration of the total electron density within each basin gives the valence electron counts of 2.49 electrons for the lone pair-like attractor and 1.76 electrons for the Ge−Ge bond attractors. Thus, the total valence electron count for valence shell basin sets is 5.13 electrons per Ge atom [$= 2.49e^-$ ("lone pair") $+ 3 \times 1.76e^-/2$ ("bond pairs")], which can be written as $Ge^{1.13-}$. As a result, the bonding situation in $EuGe_2$ can be described as $Eu^{2.26+}[Ge^{1.13-}]_2$, which agrees reasonably well with a Zintl–Klemm representation of $Eu^{2+}[Ge^-]_2$.

For ternary Eu–Zn–Ge phases, the relative electronegativities suggest that Ge will attract electrons more strongly than Zn atoms. Thus, as Ge replaces Zn in $Eu(Zn_{1-x}Ge_x)_2$ and creates no short Ge−Ge contacts, the formal charge at Ge is −4, and the formal oxidation state of Zn will change from −1 in $EuZn_2$ to +2 in EuZnGe. Experimental determination of the minimum Ge content for AlB_2-type single phases is $EuZn_{1.00}Ge_{1.00(2)}$ [$x = 0.50(1)$; Fig. 7]. According to the Zintl–Klemm formalism, there can be no short Ge−Ge contacts within the planar 6^3 nets, so the only solution is an alternant $^2_\infty[ZnGe]^{2-}$ net (Fig. 9). As a corollary, in AlB_2-type EuZnGe, there will also be no short Zn−Zn contacts. Since X-ray diffraction

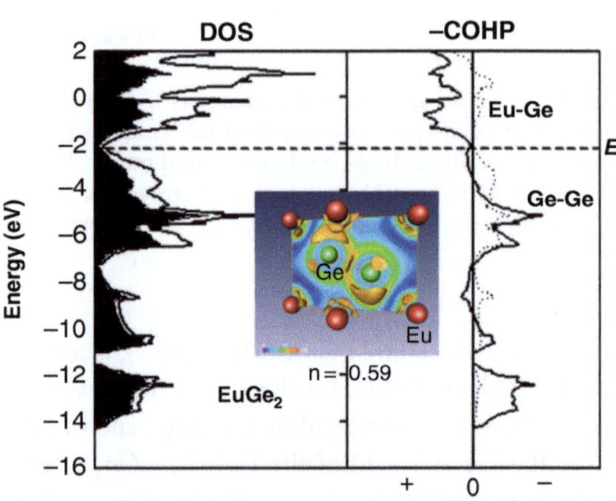

Fig. 8 DOS, COHP curves, and ELF map for $EuGe_2$. (Left) Total DOS (solid line), Eu PDOS (white region), and Ge PDOS (black region). (Right) Eu–Ge and Ge–Ge COHP curves. E_F for $10e^-$ $EuGe_2$ is indicated by the dashed line. (Inset) ELF map of $EuGe_2$

cannot distinguish Zn and Ge atoms, a completely random distribution of these two atoms is, indeed, possible, but would allow for homonuclear Zn–Zn and Ge–Ge contacts, which will violate the Zintl–Klemm formalism. Random distribution of Zn and Ge atoms within the 6^3 nets creates a binomial distribution of four different nearest neighbor environments at each Ge site:

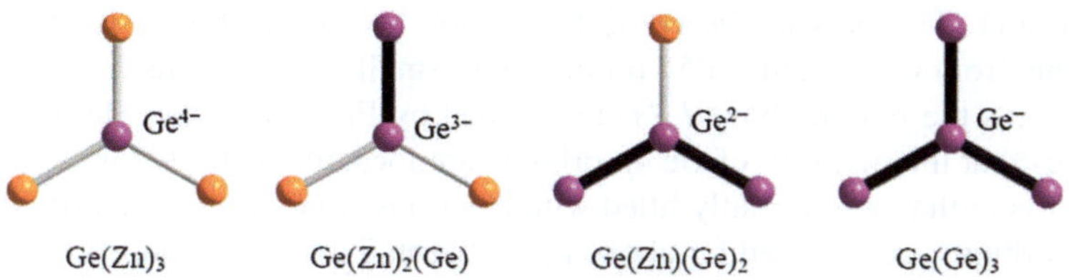

In EuZnGe, this model gives an average formal charge at Ge of −2.5, which implies 1.5-bonded Ge atoms, on average, to other Ge atoms. Likewise, the average formal charge at Zn is +0.5. This evaluation of the local electronic structure at Ge, as well as Zn, contradicts expectations from electronegativity arguments as well as electronic structure theory.

To investigate this question, TB-LMTO calculations were performed on ten structural models of EuZnGe, constructed to account for the different nearest and second nearest neighbor contacts among the Zn and Ge atoms; nearest neighbor contacts are within each 6^3 net, second nearest neighbor contacts are between adjacent 6^3 nets [147]. Among these models, the lowest energy was calculated for

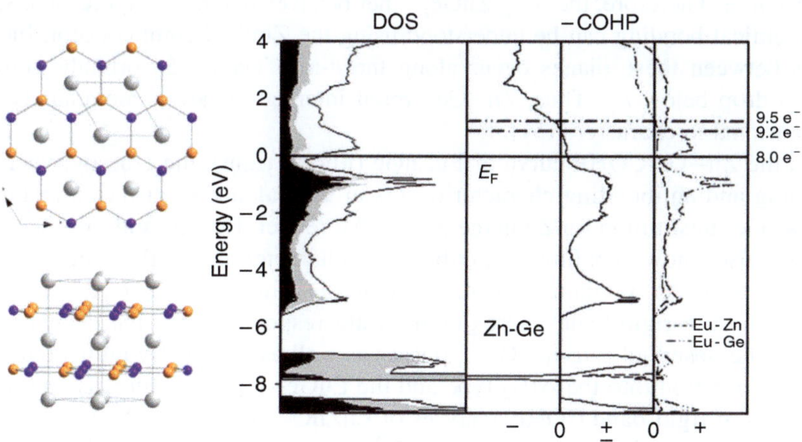

Fig. 9 $Eu(Zn_{0.5}Ge_{0.5})_2$: Lowest energy structure, DOS and COHP curves. (Right) Total DOS (solid line), Eu PDOS (white region), Zn PDOS (gray region), and Ge PDOS (black region). (Right) Zn–Ge, Eu–Zn and Eu–Ge COHP curves. E_F (solid line) is the energetic reference (0 eV). The numbers of valence electrons, which correspond to local minima in the TDOS curve as well as optima of the COHP curves, are also shown

the model containing only heteroatomic Zn–Ge and Zn···Ge interactions, which involves two $^{2}_{\infty}[ZnGe]^{2-}$ 6^3 sheets per unit cell, space group *P*6$_3$/*mmc* (see also Fig. 9). Thus, electronic structure calculations on EuZnGe imply the formulation $Eu^{2+}Zn^{2+}Ge^{4-}$. Figure 9 also displays DOS and COHP curves of this compound, where E_F is located slightly above the minimum of a pseudogap in the DOS. Throughout the entire DOS curve, significant mixing between valence orbitals of Eu, Zn, and Ge atoms is observed. The region below E_F shows significant contributions from Ge 4*p* and Eu 5*d* orbitals with small contributions from Zn 4*s* and 4*p*, whereas the region above E_F is dominated by Eu 5*d* orbitals. The DOS curve indicates that the majority of Ge 4*p* orbitals are located mostly within 5 eV below E_F, implying they are formally filled with electrons, whereas the majority of Zn 4*s* and 4*p* orbitals are centered far above E_F, between 7 and 15 eV above E_F, suggesting they are formally empty.

The experimental Zn–Ge bond distance of 2.5190(3) Å in the hexagonal nets [146], a distance which agrees well with the sum of their covalent radii (2.47 Å: *r*(Zn) = 1.25 Å, *r*(Ge) = 1.22 Å [149]), and is less than the average Zn–Zn and Ge–Ge distances in $EuZn_2$ (2.699 Å) [134] and $EuGe_2$ (2.564 Å) [148], indicates strong polar covalent interactions, which can be attributed to participation of both σ-bonding interactions and π-bonding interactions. Zn–Ge σ-bonding bands span between ca. 6 and 0.2 eV below E_F, while π-bonding bands start ca. 4.2 eV below E_F. The importance of these Zn–Ge π-interactions is reinforced when compared to the computational results on a 4b–3D "EuZnGe" in the hexagonal LiGaGe-type structure [150]: with respect to the preferred model (Fig. 9), this 4b–3D hypothetical structure is ca. 1.00 eV per formula unit higher in energy, although Zn–Ge orbital interactions are optimized and the Fermi level lies in a pseudogap in the DOS curve. Therefore, the 6^3 $^{2}_{\infty}[ZnGe]^{2-}$ net behaves like $8e^-$ graphite layers, and its chemical bonding can be understood using the Zintl–Klemm concept. Interactions between these planes occur along the *c*-axis via Eu 5*d* orbitals, some of which drop below E_F. Thus, Zn–Ge orbital interactions are predominantly two-dimensional in character [151].

In the Zn–Ge COHP curve of EuZnGe (Fig. 9, right), the crossover between bonding and antibonding character occurs at 9.2 valence electrons, a result that allows Ge substitution for Zn in the 6^3 nets. Moreover, Eu–Ge and Eu–Zn COHP curves also show bonding or nonbonding character up to this energy value. Therefore, these predicted orbital interactions in $EuZn_{1-y}Ge_{1+y}$, based on EuZnGe via a rigid-band approximation, are responsible for the decreasing *a*-axis in the observed $EuZn_{1-y}Ge_{1+y}$ phases as well as eventually influential to the phase separation into the AlB_2-type and the $EuGe_2$-type structure. Therefore, on the basis of rigid-band COHP analyses of EuZnGe, the expected upper bound of AlB_2-type phases is approximately $Eu(Zn_{0.20}Ge_{0.80})_2$, which exceeds the experimental result of $Eu(Zn_{0.30(2)}Ge_{0.70})_2$ containing 8.8 valence electrons [146]. This disagreement with the observed upper bound may be due to the limitation of the rigid-band approximation applied in this analysis, which was already apparent for the Zn-rich phases. Therefore, a Ge-rich model, $Eu(Zn_{0.25}Ge_{0.75})_2$, was studied theoretically.

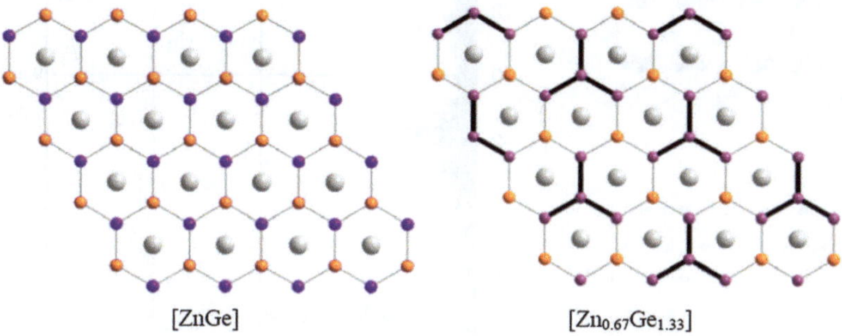

Before addressing the theoretical results, let us apply the Zintl–Klemm formalism, which should be compatible with the structural constraints in this $EuZn_{1-y}Ge_{1+y}$ region. Based on the computational results for EuZnGe, the most likely scenario is random replacement of Zn by Ge on the alternant 6^3 net, a solution that guarantees no close Zn–Zn contacts in the 6^3 net. Furthermore, as the Ge content increases, the connectivity of Ge atoms steadily increases. At low concentrations (small *y* values), trigonal Ge_4 stars emerge. For these fragments to remain planar, their formal charge should be −8: $[Ge_4]^{8-}$ would be isoelectronic to the planar $CO_3{}^{2-}$ ion, and there are numerous examples of these Zintl ions [152]. On the other hand, simply from the Ge–Ge connectivity, the formal charge can be −10, but we would expect the central Ge atom to adopt a pyramidal local coordination because $[Ge_4]^{10-}$ is isoelectronic with $SO_3{}^{2-}$. During the structural refinements of AlB_2-type $EuZn_{1-y}Ge_{1+y}$ phases [146], the U_{33}/U_{11} ratios at the (Zn/Ge) sites distinctly increases as the Ge content increases, an observation which indicates increasing tendency toward puckering of the 6^3 nets. This ratio is 1.9 in $Eu(Zn_{0.5}Ge_{0.5})_2$ and increases to 7.0 in $Eu(Zn_{0.30(2)}Ge_{0.70})_2$; all refinement attempts converged toward planar 6^3 nets, even when space groups were selected that allowed the *z*-parameter of the (Zn/Ge) sites to be unrestricted. In these models, if the Zn atoms are kept widely separated, all Ge atoms will belong to Ge_4 stars in the limiting case, $Eu_3Zn_2Ge_4 = EuZn_{0.67}Ge_{1.33} = Eu(Zn_{0.33}Ge_{0.67})_2$, which agrees nicely with the experimental upper bound of the AlB_2-region. Further replacement of Zn atoms by Ge atoms would create larger Ge-based fragments that would grow continuously until $EuGe_2$ is achieved; this feature does not occur experimentally.

The structural model created for the 9e$^-$ $Eu(Zn_{0.25}Ge_{0.75})_2$ was constructed to minimize the numbers of homoatomic interactions; this model is topologically distinct from the Ge substitution pattern above because the Zn atoms occupy *ortho*-positions in the six-membered rings in $Eu(Zn_{0.25}Ge_{0.75})_2$, but *meta*-positions in the Ge-rich models above, e.g., $Eu(Zn_{0.33}Ge_{0.67})_2$. Figure 10 illustrates DOS and nearest neighbor COHP curves for this model. Given the Ge-rich composition, the Ge PDOS is more pronounced than the Zn PDOS throughout the entire DOS curve. A pseudogap, which was clearly observed just below E_F in

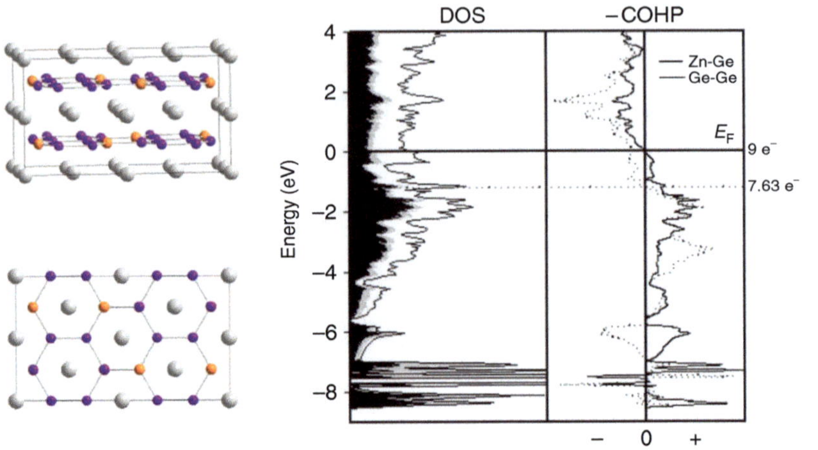

Fig. 10 Structure, DOS and COHP curves for AlB_2-derived $Eu(Zn_{0.25}Ge_{0.75})_2$. (Left) Structure projected along *a*-and *c*-directions; (Right) Total DOS (solid line), Eu PDOS (white region), Zn PDOS (gray region), and Ge PDOS (black region) as well as Zn–Ge (solid line) and Ge–Ge (dotted line) COHP curves. E_F (solid line) is the energetic reference (0 eV). The numbers of valence electrons for two optima in the COHP curves are also shown

the DOS diagram of $Eu(Zn_{0.5}Ge_{0.5})_2$ (Fig. 9), is no longer noticeable because of the more pronounced contributions from Eu 5*d* and Ge 4*p* orbitals near E_F. The Zn−Ge COHP curve is fully optimized at E_F, while the Ge−Ge COHP curve falls well within a strong antibonding region; the Ge−Ge COHP curve is optimized at ca. 7.63e$^-$. Thus, the change in composition from EuZnGe to $Eu(Zn_{0.25}Ge_{0.75})_2$ significantly affects the calculated DOS and predicts a lower upper bound for the AlB_2-type structures than suggested by analysis of the DOS and COHP curves for EuZnGe. According to the COHP analysis of $Eu(Zn_{0.25}Ge_{0.75})_2$, moreover, planar 6^3 nets are not especially favorable environments to accommodate the Ge−Ge orbital interactions. Thus, the crystal structure undergoes a structural distortion as well as phase separation to remove this instability. In $Eu(Zn_{0.25(2)}Ge_{0.75(2)})_2$, an incommensurately modulated structure along the *c*-direction was observed, and its solution relied on 3+1 D super-space analysis with twinning, shown in Fig. 11 [146]. The resulting structural model showed puckering of the 6^3 nets, which is rather constant and changes direction at regular intervals along the stacking direction. Presumably, further details of the model are influenced by how Zn and Ge are distributed throughout the entire arrangement, but this effect could not be solved by X-ray diffraction results.

To investigate puckered vs. planar hexagonal layers in $Eu(Zn_{0.25}Ge_{0.75})_2$, the total energies of five structurally relaxed models were explored [147] using VASP [153–156]. In four of these models, each hexagonal layer had a fixed composition $[Zn_{0.5}Ge_{1.5}]$ with the Zn atoms completely separated from one another. The fifth model consisted of a 1:1 intergrowth structure of EuZnGe and $EuGe_2$. The two models giving the lowest total energy after structural relaxation are shown in

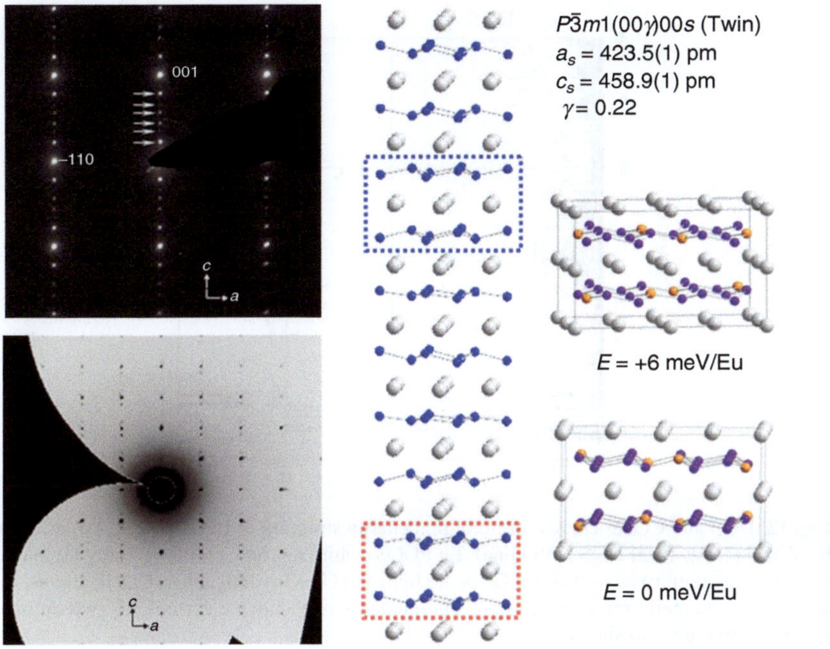

Fig. 11 $Eu(Zn_{0.25}Ge_{0.75})_2$ (Left, top) HRTEM of (110) plane, incommensurately modulated peaks are marked with arrows. (Left, bottom) Reciprocal space reconstructed from X-ray diffraction data displayed along the (110) plane. (Middle) Incommensurately modulated structures of $Eu(Zn_{0.25}Ge_{0.75(2)})_2$ saw-tooth like modulations. (Right) Two model structures of $Eu(Zn_{0.25}Ge_{0.75})_2$ after relaxation using VASP. Two local environments of puckered layers resembling these two models are highlighted in the incommensurately modulated structure

Fig. 11: the model with out-of-phase puckered layers is the most favorable structure, but one with in-phase puckered layers is just 6 meV/f.u. higher in energy. In these models, the angular deficiencies of the sum of the three bond angles from 360° at each atomic site, which is a measure of pyramidalization, are 16.2° for the Zn-site, and 15.4° and 17.7° for the Ge-sites. These values compare very favorably with the range 8.2°–18.4° calculated from the incommensurately modulated solution [146]. The energetic similarity between these two models substantiates the super space solution for this incommensurately modulated structure. Puckering of each 6^3 net is intrinsic with the Zn and Ge atoms uniformly distributed throughout each network. The intergrowth model relaxed toward planar [ZnGe] and $[Ge_2]$ nets and an energy value ca. 370 meV/f.u. above the lowest energy structure, a computational results which discounts this model as a reasonable solution to the structure of $Eu(Zn_{0.25}Ge_{0.75(2)})_2$ [146].

The effect of puckering on the DOS as well as Zn–Ge and Ge–Ge COHP curves is presented in Fig. 12, which shows these curves based on a TB-LMTO-ASA calculation of the lowest energy model obtained by VASP, and should be compared

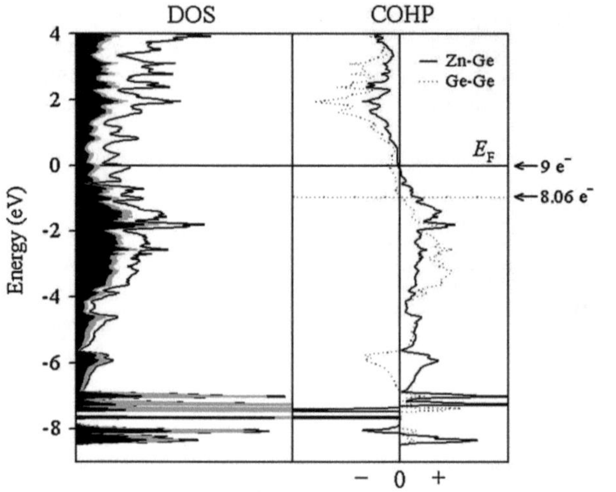

Fig. 12 DOS and COHP curves for minimum energy structure of $Eu(Zn_{0.25}Ge_{0.75})_2$ as determined by VASP [153]. Total DOS (solid line), Eu PDOS (white region), Zn PDOS (gray region), and Ge PDOS (black region) as well as Zn–Ge (solid line) and Ge–Ge (dotted line) COHP curves. E_F (solid line) is the energetic reference (0 eV). The numbers of valence electrons for two optima in the COHP curves are also shown

against Fig. 10. At $9e^-$, Zn–Ge orbital interactions are optimized, but there are two significant differences between the two figures that favors the puckered model: (1) E_F falls in a narrow pseudogap of the DOS of the puckered model as compared to the planar case; and (2) the Ge–Ge COHP curve is essentially nonbonding in the puckered case, where it is distinctly antibonding in the planar case. Both of these features in the electronic structure contribute to the preference toward puckering of the 6^3 nets in $Eu(Zn_{0.25}Ge_{0.75(2)})_2$.

Thus, along the $Eu(Zn_{1-x}Ge_x)_2$ series, the Zintl–Klemm formalism, which relates local electronic structure and chemical bonding features, changes validity at EuZnGe. $EuZn_2$ is "electron deficient" and does not satisfy the Zintl–Klemm formalism: the 4b–3D network of Zn atoms would require four valence electrons per Zn atom. Nevertheless, Zn is more electronegative than Eu, and the formal oxidation state of Zn is -1, a state which assigns just three valence electrons per Zn atom. However, in a metal there can be no charge build-up. $EuGe_2$ is "electron precise" and does satisfy the Zintl–Klemm formalism: the 3b–2D network of Ge atoms requires three valence electrons per Ge atom. Again, the formal oxidation state of Ge is -1, a state which assigns five valence electrons per Ge atom. In ternary $EuZn_{1-y}Ge_{1+y}$ phases, Ge will attract electrons more strongly than Zn atoms. Although no indication of long-range ordering of Zn and Ge can be detected by X-ray diffraction, a certain degree of short-range ordering is suggested by electronic structure theory. At present, however, there is no experimental corroboration of this suggestion.

3.3 The $Eu(Ga_{1-x}Ge_x)_2$ System

Experimental results indicate some similarities as well as distinct differences between $Eu(Zn_{1-x}Ge_x)_2$ and $Eu(Ga_{1-x}Ge_x)_2$, results which are partially coupled to the number of valence electrons [157]. Hexagonal AlB_2-type phases exist for $x < 0.50$; although no precise upper or lower boundaries of this phase region are specified; $EuGa_2$ adopts the orthorhombic KHg_2-type structure, and there is certainly a homogeneity width between at least $x = 0.25$ and ca. 0.40. As in Eu $(Zn_{1-x}Ge_x)_2$, no near neighbor Ge–Ge contacts are assumed to exist in the 6^3 nets of $Eu(Ga_{1-x}Ge_x)_2$, and the U_{33}/U_{11} ratio clearly increases as the Ge content increases toward the $9e^-$ EuGaGe. The crystallographic results for $Eu(Ga_{0.55}Ge_{0.45(2)})_2$ suggested puckered 6^3 nets [157]. These structural observations are in agreement with a Ga–Ge COHP analysis of an AlB_2-type "EuGaGe," an analysis which showed optimum Ga–Ge orbital interactions at ca. $8.25e^-$ and clearly showed Ga–Ge antibonding states above $8.5e^-$ arising from in-plane Ga–Ge σ^* overlap. Therefore, the homogeneity range for AlB_2-type $Eu(Ga_{1-x}Ge_x)_2$ is predicted to be $0.12 < x < 0.50$, which means between $8.25e^-$ and $9e^-$, which agrees well with experiment.

For $x \geq 0.50$, three ternary examples are $x = 0.50$, 0.55, and 0.60, all of which may be formulated as $[EuGaGe]_u[EuGe_2]_v$ to emphasize the intergrowth structures detected in the $x = 0.55$ and 0.60 phases (Fig. 13). The YPtAs-type structure of EuGaGe shows puckered 6^3 nets with two adjacent nets puckered out-of-phase to

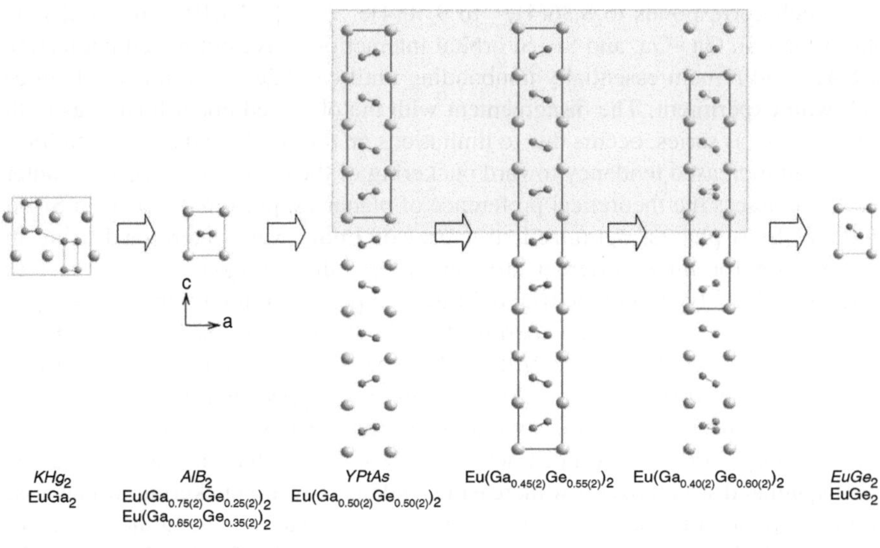

Fig. 13 Crystal structures along the $Eu(Ga_{1-x}Ge_x)_2$ ($0 \leq x \leq 1$) series. The structure types are given in italics. The shaded region emphasizes the YPtAs-type building blocks contained in the three phases EuGaGe, $EuGa_{0.9}Ge_{1.1}$, and $EuGa_{0.8}Ge_{1.2}$

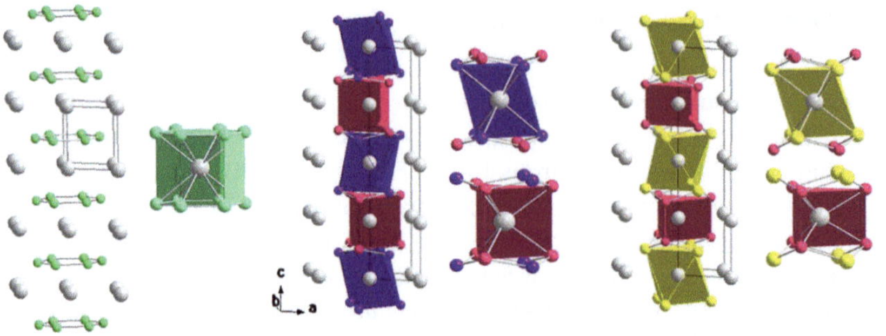

Fig. 14 Crystal structures of EuGaSi, EuGaGe and EuGaSn and coordination environments at the Eu atoms. Eu: large light gray; Ga/Si: small light gray/green; Ga: small dark gray/red; Ge: small black/purple; Sn: small white/yellow

create one shorter and one longer interplanar spacing distance. Based on a similar structure for EuGaSn [158], shown in Fig. 14, in which there is greater contrast between Ga and Sn atoms, each 6^3 net contains alternating Ga and Ge atoms with the shorter interplanar spacing between adjacent Ga atoms, although this Ga···Ga distances remains nonbonding, i.e., 3.726(1) Å. The corresponding distance in EuGaSn is 3.178(1) Å. On the other hand, EuGaSi adopts an AlB_2-type structure with Ga and Si disordered within 6^3 nets and an interplanar spacing of 4.554(1) Å [158]. In fact, the stability range of AlB_2-type $Eu(Ga_{1-x}Si_x)_2$ is $0.18(2) \leq x \leq 0.70$ (2), which corresponds to 8.36(4)e^- to 9.40(4)e^- [157]. COHP analysis of intraplanar Ga–Si, Ga–Ga, and Si–Si orbital interactions gave optimized interactions at 8.4e^- and remain essentially nonbonding until ca. 9.2e^-, a result which agrees well with experiment. The disagreement with the observed upper bound, as in the $Eu(Zn_{1-x}Ge_x)_2$ series, occurs due to limitations on the rigid-band approximation as well as an increased tendency toward puckering of the 6^3 nets for larger Si content [157]. To assess the theoretical preference of planar vs. puckered 6^3 nets in Si-rich $Eu(Ga_{1-x}Si_x)_2$ phases, optimized structures of $EuSi_2$ were determined using the VASP code for three different structure types: the tetragonal $ThSi_2$ type; the hexagonal AlB_2 type; and the trigonal $EuGe_2$ type. Accordingly, the $ThSi_2$ type is preferred, in agreement with experiment [159], by 61 meV/$EuSi_2$ over the $EuGe_2$ type, and by 164 meV/$EuSi_2$ over the AlB_2 type. These theoretical results strongly suggest that the Si-rich $Eu(Ga_{1-x}Si_x)_2$ phases contain puckered 6^3 nets.

The differences between the structural and phase behavior of $Eu(Ga_{1-x}Si_x)_2$ and $Eu(Ga_{1-x}Ge_x)_2$ are subtle yet distinct. As the tetrelide content increases, the 6^3 nets in AlB_2-related structures show increasing tendency toward puckering, which arises from occupation of π-antibonding orbitals. The binary tetrelides, $EuSi_2$ and $EuGe_2$, show two different modes of disrupting this interaction [127, 160]. In $EuSi_2$, all Si atoms are locally trigonally planar by three Si atoms, but the planes of adjacent Si atoms are perpendicular to each other. On the other hand, all Ge atoms in $EuGe_2$ are trigonally pyramidally coordinated by three Ge atoms to create puckered 6^3 nets.

Both of these observations suggest that under slightly reduced environments, Si prefers trigonal planar coordination while Ge prefers tetrahedral coordination:

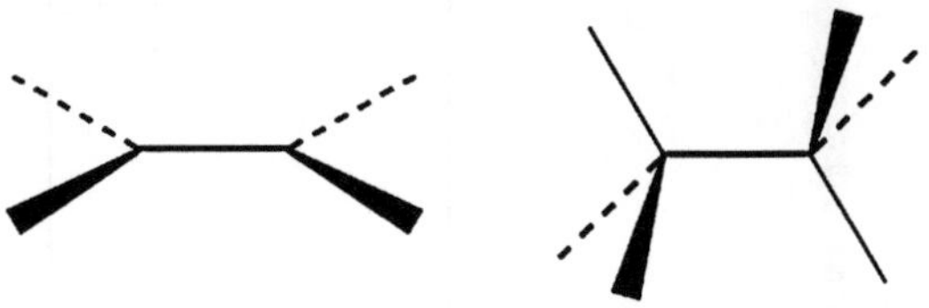

A semiquantitative analysis of Ga$-$*Tt* interactions considering trigonal planar vs. tetrahedral coordination indicates that at 9e$^-$, local trigonal planar coordination is favored for the silicides whereas local tetrahedral coordination is preferred for the germanides. For 8e$^-$, tetrahedral coordination is preferred for both silicides and germanide systems [157].

The DOS and COHP curves of EuGaGe in Fig. 15 show a moderate pseudogap near E_F, with substantial contribution from Eu 5*d* bands. The Ga$-$Ge COHP curve is optimized at 8.50 valence electrons (ca 0.6 eV below E_F) and remains essentially nonbonding up to ca. 9.25 valence electrons (ca. 0.3 eV above E_F); meanwhile, the interplanar Ga···Ga COHP curve shows significantly weaker orbital interactions except for a σ-bonding contribution just below E_F, which intersects the crossing between bonding states below and antibonding Ga···Ga states above this energy value. Computational studies invoking ELF analysis indicated that the observed puckering mode in EuGaGe is a result from the weakly attractive orbital interaction between interplanar Ga atoms and strongly mediated by multicenter orbital interactions between Ga 4*s* and 4*p* orbitals with Eu 5*d* orbitals [158].

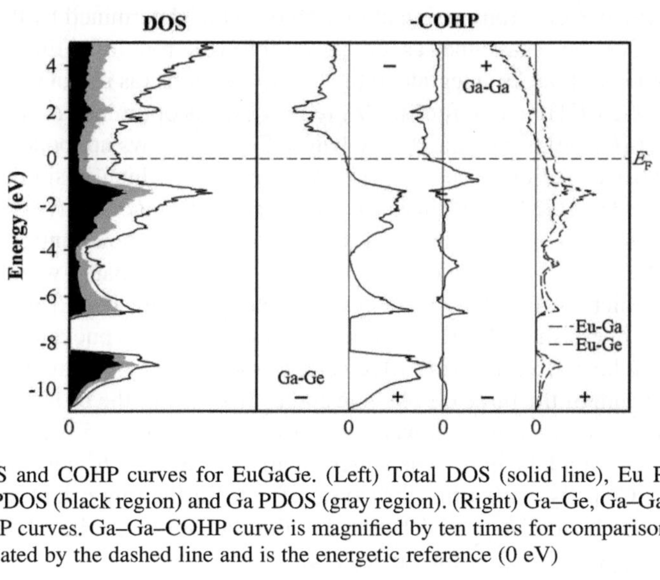

Fig. 15 DOS and COHP curves for EuGaGe. (Left) Total DOS (solid line), Eu PDOS (white region), Ge PDOS (black region) and Ga PDOS (gray region). (Right) Ga–Ge, Ga–Ga, Eu–Ga and Eu–Ge COHP curves. Ga–Ga–COHP curve is magnified by ten times for comparison. The Fermi level is indicated by the dashed line and is the energetic reference (0 eV)

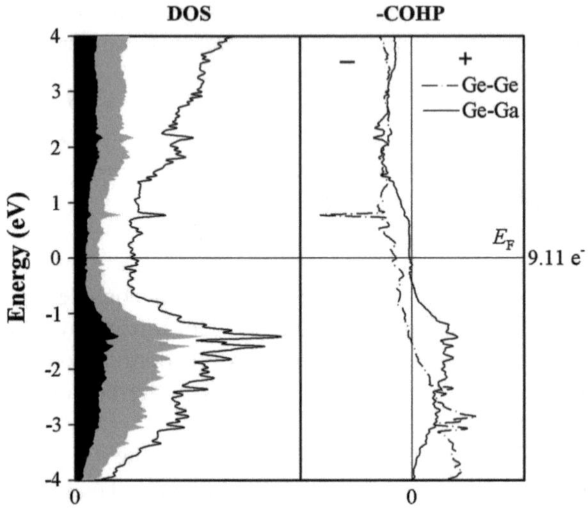

Fig. 16 DOS and COHP curves for $Eu(Ga_{0.45}Ge_{0.55})_2$. (Left) Total DOS (solid line), Eu PDOS (white region), Ge PDOS (gray region), and Ga PDOS (black region). (Right) Ga–Ge, Eu–Ga, and Eu–Ge COHP curves. The Fermi level is indicated by the solid line

$Eu(Ga_{0.45}Ge_{0.55(2)})_2$ contains 8 [GaGe] and 1 [Ge_2] puckered 6^3 nets, and given this description, its DOS curve, shown in Fig. 16 within 4 eV of E_F, is similar to the DOS curve of EuGaGe [158]. The strongest peak occurs at ca. 1.7 eV below E_F, while there are no apparent pseudogaps. There is more Ge 4*p* orbital contribution to the total DOS than Eu 5*d* orbitals or Ga 4*p* orbitals below E_F because of the Ge-rich composition and relative electronegativities, while Eu 5*d* orbital contributions become greater near and above E_F. The intergrowth character of this structure is validated by evaluating the number of valence electrons assigned to each 6^3 net as determined by the integrated total DOS curve: the net assigned as [Ge_2] gives 9.98 electrons (as in $EuGe_2$), while the others assigned as [GaGe] integrate to 8.98–9.01 electrons (as in EuGaGe).

The Ga−Ge COHP curve for $Eu(Ga_{0.45}Ge_{0.55(2)})_2$ is optimized at ca. 8.6 valence electrons per [GaGe] 6^3 net (ca. 0.5 eV below E_F) and shows nonbonding character up to ca. 9.6 valence electrons (ca. 0.7 eV above E_F), which is similar to that of EuGaGe. The Ge−Ge COHP curve for the puckered hexagonal layer consisting of entirely Ge atoms is optimized at ca. 1.4 eV below E_F and shows antibonding states at E_F. This Ge-only layer in $Eu(Ga_{0.45}Ge_{0.55(2)})_2$ can be compared with Ge layers in $EuGe_2$, for which the Ge−Ge COHP curve is nearly optimized at E_F with $10e^-$. The difference between these COHP curves arises from different puckering geometries of the two Ge layers, a puckering which can be measured by the angular deficiency between the sum of the three Ge–Ge–Ge angles from 360°: the 6^3 [Ge_2] layer in Eu $(Ga_{0.45}Ge_{0.55(2)})_2$ has a much lower angular deficiency of 17.5° at each Ge atom than in $EuGe_2$, which is 39.0°. The wider separation between 6^3 nets in $EuGe_2$ than in $Eu(Ga_{0.45}Ge_{0.55})_2$ causes this structural effect. Furthermore, and as a result, a shorter Ge−Ge bond distance is observed in $Eu(Ga_{0.45}Ge_{0.55})_2$, i.e., 2.527 Å for Eu $(Ga_{0.45}Ge_{0.55})_2$ and 2.552 Å for $EuGe_2$ [148].

Thus, the Zintl–Klemm formalism assisted in the interpretation of phase and structural behavior among the $Eu(Zn_{1-x}Ga_x)_2$, $Eu(Zn_{1-x}Ge_x)_2$, and $Eu(Ga_{1-x}Ge_x)_2$ systems, but these interpretations were mediated by electronic structure calculations. The effects of electronegativity on the relative atomic orbital energies make a rigid-band approximation of limited utility when predicting phase behavior based solely on electron count, even though size effects are minimal in each of these three systems. The instability of the planar 6^3 net for $9e^-$ is manifested by puckering in both the $Eu(Zn_{1-x}Ge_x)_2$ and $Eu(Ga_{1-x}Ge_x)_2$ systems, although with different overall results; this behavior is related to, yet distinctly different from, the $9e^-$ *AeTrTt* superconductors [114–120]. Indeed, part of the driving force toward puckering involves multicentered bonding with the 5*d* orbitals of the Eu atoms, in addition to the infiltration of occupied π^* orbitals in the DOS. In the subsequent section, we will examine the size effects of the "cations" in Zintl phases.

4 Ionicity, Covalency, and Metallicity: The NaTl-Type Revisited

As mentioned in Sect. 1, the Zintl–Klemm formalism was established assuming the simultaneous presence of ionic and covalent bonding characteristics in Zintl phases. Valence electrons are transferred from electropositive metals to electronegative (semi)metals, resulting in "cations" and "anionic" frameworks. The bonding within the frameworks is considered as largely *covalent*, and, thus, their structures are rationalized through electron counting rules, viz., the octet rule [1] or Wade–Mingos rules [5, 6]. Although the coexistence of ionic and covalent interactions is assumed in the application of the Zintl–Klemm concept, the covalent portion has been emphasized much more than the ionic portion. In fact, the "cations" are often treated as mere electron donors, charge balancers, and by-standers in constructing the structures of Zintl phases. The formal ignorance of these "cations" contributes to the idea of *pseudoatoms*, that is, after gaining extra valence electrons, the electronegative atoms will behave like isoelectronic neutral atoms. For example, Tl^- is isoelectronic with a tetrelide element; hence, it adopts a diamond structure in NaTl [19].

The treatment of these "cations" as mere observers is an oversimplification and can limit the applicability of the Zintl–Klemm formalism. For instance, while Tl^- can be perceived as a pseudo-tetrel atom in NaTl, a similar rationalization is unclear in LiTl [78] and KTl [161] (Fig. 17) – the Tl networks in these two compounds bear no resemblance to any solid-state or solution-phase tetrel structure. LiTl is isosteric with CsCl, which is a structure that can be understood, intuitively, through *ionicity* or *metallicity* better than *covalency*. In KTl, Tl atoms form distorted octahedra, formally $[Tl_6]^{6-}$, which has C_{2v} site symmetry. Although Tl can still be perceived as four-bonded in KTl, it is significantly different from any tetrel element. We should point out, however, that gas-phase Si_6 clusters correspond to one of the "magic numbers" of Si-based clusters [162, 163]; according to quantum mechanical calculations, the preferred structure is a distorted octahedron, adopting possible

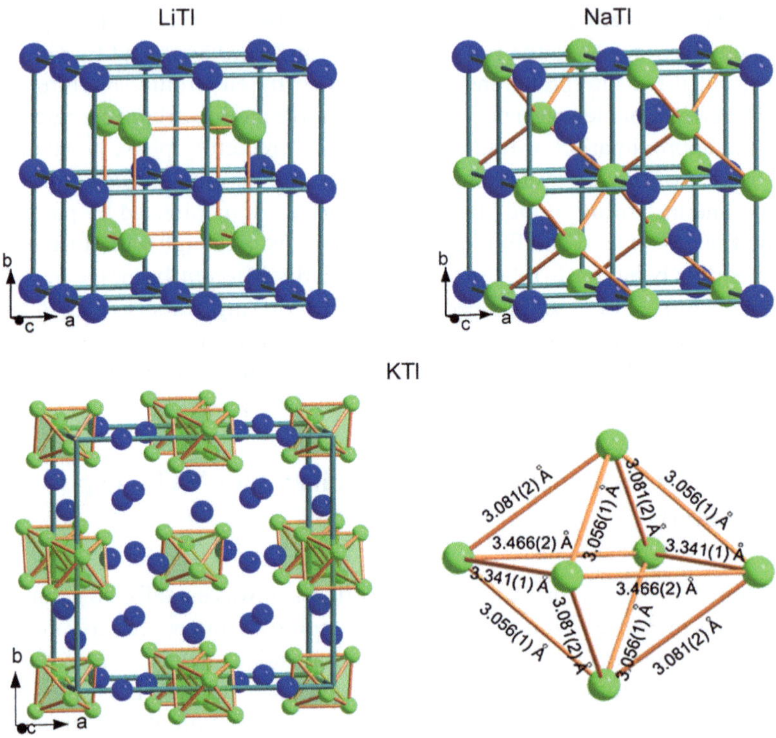

Fig. 17 Crystal structures of LiTl, NaTl, and KTl (black/blue: Li, Na, and K; gray/green: Tl)

point groups C_s, C_{2v}, or D_{4h} [164, 165]. For cases like KTl, the Zintl–Klemm formalism can remain applicable by dropping the *pseudoatom* concept strictly, retaining only the octet rule [8]. Thus, Tl^- does not have to form a known tetrel structure, but the average number of bonds at every Tl^- should be four. After this tailoring of the simple bonding concept, both NaTl and KTl follow the Zintl–Klemm formalism; but the difference between NaTl and KTl still cannot be rationalized, which apparently stems from the difference in "cations."

Besides *covalent* and *ionic* interactions, *metallic* behavior can also be present in Zintl phases. According to the DOS and COHP curves of NaTl (Fig. 18), calculated on the experimental structure using TB-LMTO, the majority of the states below the Fermi level (E_F) arise from Tl 6*s* and 6*p* orbitals, although Na atomic orbitals also contribute significantly, especially within 4 eV of E_F. Thus, the *ionic* picture is not precise. Moreover, E_F is located right at the bonding–antibonding crossover of the Tl–Tl COHP curve, which indicates optimization of the Tl–Tl *covalent* interaction. The integrated Tl–Tl COHP value, which is a measure of covalent bond index, is −0.871 eV, and significantly larger than the integrated Na–Tl and Na–Na COHP values, respectively, −0.514 and −0.241 eV, although the three nearest neighbor distances are all the same. Nevertheless, there is no band gap in the DOS

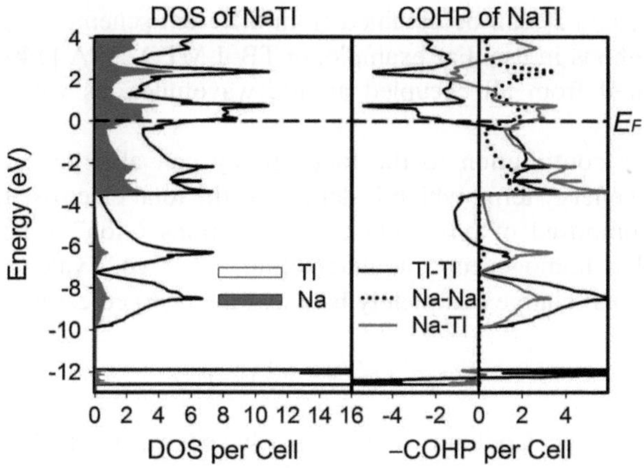

Fig. 18 DOS and COHP curves of NaTl

curve at E_F; a "pseudogap" is present just below E_F. Therefore, NaTl is predicted to be metallic, neither insulating like diamond nor semiconducting like Si; the electrical conductivity of NaTl at 20°C is $1.23 \times 10^{-4}/\Omega$ cm and it decreases with increasing temperature [166]. Therefore, in a Zintl phase, there can be an intermingling of *ionic*, *covalent*, and *metallic* characteristics. The originality of the Zintl–Klemm concept lies in the consideration of *covalency* in intermetallic compounds. This does not mean that one should overlook or oversimplify the *ionic* and *metallic* pictures. To better understand composition–structure relationships of Zintl phases, we should evaluate the *covalent* interaction in combination with the *ionic* and *metallic* contributions.

The effects of both ionic and metallic interactions on structure can be approximated by an electrostatic energy or Madelung energy, but with different details. To evaluate the *ionic* effect, we can calculate the lattice energy, U_L, according to the Born–Mayer equation [167]:

$$U_L = \frac{M}{4\pi\varepsilon_0}\frac{z_1 z_2 e^2}{r_0}\left(1 - \frac{\rho}{r_0}\right), \tag{1}$$

where z_1 and z_2 are the charges of the cation and anion, ε_0 is the vacuum permittivity, r_0 is the nearest neighbor distance between neighboring cation and anion, M is the Madelung constant, and ρ is a parameter reflecting the repulsion between adjacent electron clouds. For comparisons between different structures, however, the repulsive ρ/r_0 term may be dropped, which leaves just the first term, called the Madelung energy, or Madelung part of the lattice energy, MAPLE [168, 169]. Values of M have been tabulated for many structures [170] and can be calculated for any new structure by an Ewald summation [171]. Because of typically incomplete charge transfer between atoms of different electronegativities, the challenge is to assign partial charges z_1 and z_2 to "cations" and "anions." In quantum mechanical calculations, z_1 and z_2 can be obtained from different schemes depending on the calculation methods in use. For example, in TB–LMTO–ASA [140], it can be the charge integrated from all occupied atomic wavefunctions within each atomic sphere.

The *metallic* contribution to the total energy can also be evaluated with a Madelung-type energy term, which is defined as the total electrostatic (ES) energy of a system composed of positively charged cations ("ion" = nuclei plus core electrons) and a homogeneous noninteracting, or "free," valence electron gas (FEG), whose total charge completely balances the charged cations [172]:

$$E_{\mathrm{ES}} = V_{\mathrm{ion-ion}} + V_{\mathrm{ion-FEG}}, \tag{2}$$

where $V_{\mathrm{ion-ion}}$ is the Coulomb repulsion between cations and $V_{\mathrm{ion-FEG}}$ is the Coulomb attraction between ions and the homogeneous free electron gas. Its calculation takes a form similar to that in the ionic definition [173]:

Elements:

$$E_{\mathrm{ES}} = \alpha_{\mathrm{M}} \frac{z^2}{R_{\mathrm{a}}} \tag{3}$$

Binary Compounds:

$$E_{\mathrm{ES}} = \frac{1}{R_{\mathrm{a}}} \left(Z_A^2 \alpha_{AA} + Z_A Z_B \alpha_{AB} + Z_B^2 \alpha_{BB} \right), \tag{4}$$

where R_{a} is the average atomic radius; Z, Z_A, and Z_B are the charges of the cations, and α_M, α_{AA}, α_{AB}, and α_{BB} are Madelung constants, which can also be calculated with techniques similar to the Ewald summation [174]. Among various quantum mechanical computational techniques, those using a "pseudopotential" [175], e.g., VASP [153], can conveniently calculate the metallic electrostatic energy.

Due to the significant difference between the *ionic* and *metallic* definitions of the electrostatic energy, these energetic terms cannot be calculated simultaneously. This means that the contribution of *covalency* to crystal structure cannot be unambiguously singled out in a complete total energy calculation. For example, with the VASP code, we can calculate the metallic electrostatic energy and total energy. By subtracting the metallic electrostatic energy from the total energy, we obtain the electronic energy:

$$E_{\mathrm{electronic}} = T_{\mathrm{VE}} + (V_{\mathrm{ion-VE}} - V_{\mathrm{ion-FEG}}) + V_{\mathrm{VE-VE}} + E_{\mathrm{XC}}. \tag{5}$$

T_{VE} is the kinetic energy, which is not included in the electrostatic energy. Real valence electrons interact with each other, so there are interactions terms, $V_{\mathrm{VE-VE}}$ and E_{XC}, representing, respectively, the Coulomb and exchange-correlation energies between valence electrons. Moreover, the density of real valence electrons is inhomogeneous: valence electrons concentrate, or localize, at some places and

Table 3 The difference in Madelung, electronic, and total energy between the CsCl- and the NaTl-type structures, $\Delta E = E$(NaTl type)–E(CsCl type), calculated with VASP for lithium trielides. For each composition, the CsCl-type and the NaTl-type models have the same volume per formula unit, which is taken from the experimental value [78–81]

Composition:	LiAl	LiGa	LiIn	LiTl
ΔE_{ES} (eV)	+1.3064	+1.3435	+1.2202	+1.2054
$\Delta E_{electronic}$ (eV)	−1.4617	−1.4185	−1.1478	−1.0069
ΔE_{total} (eV)	−0.1553	−0.0750	+0.0724	+0.1985

deplete at others. Such inhomogeneities can be caused by covalent bonding (electrons localized between atom pairs or within clusters) as well as charge transfer (electrons localized at "anions" and depleted at "cations"). Therefore, the ($V_{ion-VE} - V_{ion-FEG}$) term, i.e., the difference of the ion-electron Coulomb potential energy between real valence electrons and the homogeneous electron gas, reflects both *covalent* and *ionic* effects, and these two features, at present, cannot be further segregated quantitatively. However, they can be evaluated qualitatively by examining the calculated valence electron density map, an analysis which has been demonstrated in Pawlowska's work [176] and our work discussed below.

Before addressing some specific examples, we must emphasize the intrinsic *isotropic* nature of *ionic* and *metallic* interactions in solids, as evaluated by these Madelung-type energy terms. Any anisotropic structural effects are condensed into the single parameter of the Madelung constant (M or α), which are obtained from the structural topology. *Covalent* interactions are necessarily *anisotropic* by building up electron density in specific regions and depleting it in others. Referring again to the van Arkel–Ketelaar diagram in Fig. 1, we see that electronegativity differences are a good measure of *ionic* character, while the electronegativity sums separate *metallic* from *covalent* character. Metallization occurs in systems whose constituent atoms have diffuse wavefunctions such that small changes in nearest-neighbor interactions do not affect the total energy greatly. Furthermore, the occupation of energy bands is low so that the driving force for a charge-density wave distortion, through either *covalent* or *ionic* interactions, is too small to overcome elastic forces of the underlying structure. There is a further dilemma: *metallic* behavior eliminates the formation of charged ions. Nonetheless, if we consider the local symmetry, compounds with greater *ionic* or *metallic* character tend to be more *isotropic* in nature; those with greater *covalent* character tend to be more *anisotropic* in nature.

4.1 The Lithium Trielides

LiAl [79], LiGa [80], and LiIn [81] are isostructural with NaTl at ambient conditions, but LiTl [78] adopts the CsCl-type structure. A complete story of the electronic structure of these lithium trielides can be obtained by combining

Christensen's [177] and Pawlowska's [176] work, both of which used the TB–LMTO method. By using a "frozen potential" approach, Christensen partitioned the total energy into a Madelung term and a band energy term. Here, the Madelung term was calculated as the electrostatic energy between partially charged atomic spheres, so it follows the *ionic* definition, and always favors the CsCl-type structure. This result is exactly expected from (1): the Madelung constants are 1.76267 for the CsCl-type structure and 1.51343 for the NaTl-type structure [170]. Hence, given the same r_0 values, and as long as the partial charges do not differ drastically between these two structure types, the CsCl type affords the lower Madelung energy.

Christensen also found that, contrary to the Madelung term, the band energy always favors the NaTl-type structure. The author claimed that this means covalent bonding favors the NaTl-type structure because the band energy "contains all the effects of bonding and hybridization" [177]. Therefore, the Madelung and the band energy terms, or *ionicity* and *covalency*, counteract one another in these phases. In LiAl, the band energy (*covalency*) wins, and in LiTl, the Madelung term (*ionicity*) wins. This is a decent rationalization of the structural differences among lithium thallides; the only discrepancy is the evaluation of *metallicity*. In fact, it is inappropriate to assign the band energy solely to *covalency*. In the calculation of Madelung energy, the charge distribution is treated as spherically symmetric within each atomic sphere. The actual charge distribution differs from this simplified picture. Electrons can be localized through covalent bonding, or they can be largely delocalized as in metals. Therefore, by subtracting the Madelung energy from the total energy, the resulting band energy should include both *covalent* and *metallic* contributions.

Pawlowska plotted the valence electron density maps for the four lithium trielides (for LiTl, the charge density was plotted with a hypothetical NaTl-type structure), maps which did confirm the presence of covalent bonding. Electron density maxima were observed clearly between Al atoms in LiAl. Such localization of valence electrons decreases from LiAl to LiTl: from Al to Ga to In, electron density maxima become less and less apparent; and in LiTl the maxima are not located between but, rather, at the Tl atoms. Therefore, from LiAl to LiTl, *covalency* is decreasing and *ionicity* is increasing, which is consistent with what we would expect from the electronegativities (Pauling's) of triels: Al, 1.61; Ga, 1.81; In, 1.78; and Tl, 2.04. Therefore, the structure switches from the *covalency*-favored NaTl type to the *ionicity*-favored CsCl type.

We reexamined Christensen's and Pawlowska's results using VASP, which uses projector augmented wave (PAW) pseudopotentials [178]. The relative total energies (Table 3) of both the NaTl- and the CsCl-type structures for LiAl, LiGa, LiIn, and LiTl using the experimental volumes per formula unit (f.u.) for both structure types reflect the observed trend in relative stability: the favored structure switches from the NaTl- to the CsCl type from Al to Tl. VASP also succeeds in predicting the right structures for LiAl, LiGa (the NaTl type), and LiTl (the CsCl type), but fails for LiIn, which is reported in the NaTl-type structure. For LiIn, VASP calculates a lower total energy for the CsCl-type structure by 72.4 meV per f.u. (36.2 meV per atom) over the NaTl-type structure. This discrepancy between experiment and theory for LiIn could be reconciled by including the zero-point vibration as well as collective vibrations to allow a theoretical assessment at room temperature; such calculations have not as yet been reported.

The total energies were also partitioned into the electronic and electrostatic (Madelung) terms, which used the metallic definition. For all of these compositions, VASP gives lower electrostatic energy for the CsCl-type structure as expected from (4): Z_A and Z_B are, respectively, +1 and +3; $E_{ES}(\text{CsCl}) = -8.16959/R_a$ and $E_{ES}(\text{NaTl}) = -8.02789/R_a$; that is, given the same volume per f.u., the CsCl-type structure gives the lower electrostatic energy. Therefore, the CsCl-type structure is favored by both *isotropic* interactions, *ionicity* and *metallicity*.

Contrary to the electrostatic energy, the electronic energy always favors the NaTl-type structure. As discussed above, both *ionic* and *covalent* effects are included in the electronic energy. However, because *ionicity* favors the CsCl-type structure, the preference of the NaTl-type structure by the electronic energy must be from *covalency*.

The valence electron density maps of the NaTl-type structure for all four trielides calculated from the VASP results (Fig. 19) agree with Pawlowska's finding [176]. The valence electrons are concentrated within cylindrical capsules along every Al–Al bond axis in LiAl, indicating covalent bonding between Al atoms. For LiGa and LiIn, electrons are still localized between triel atoms along

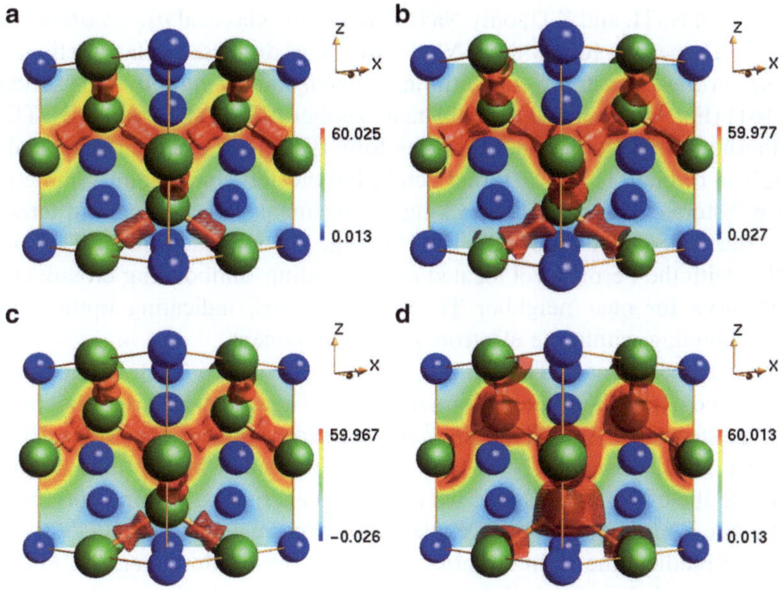

Fig. 19 Valence electron density maps of NaTl-type lithium trielides: (a) LiAl; (b) LiGa; (c) LiIn; and (d) LiTl. Dark gray/blue: Li; medium gray/green: triels. The iso-surfaces are drawn at 57 electrons per unit cell

the *Tr–Tr* bond axis, but the shapes of the capsules are no longer cylindrical but, rather, hourglass-shaped, indicating that the maxima of electron density are shifted away from the bond centers and toward the triel atoms, so that covalent bonding is weakening. In LiTl, electrons are primarily localized spherically around the Tl atoms with only slight connections between capsules, a result that reveals the diminishing of *covalency*. The weaker covalent interaction between Tl atoms can be attributed to a relativistic effect, in which the 6*s* orbitals are highly contracted and not sufficiently involved in bonding. This can be seen by comparing the band structures of LiAl and hypothetical NaTl-type LiTl [177]. For LiAl, the 3*s*–3*s* bonding states are occupied and the 3*s*–3*s* antibonding states are empty; by contrast, for NaTl-type LiTl, the 6*s*–6*s* bonding and antibonding states are both below the Fermi level. Details will be included in an impending report.

Thus, in LiAl, LiGa, LiIn, and LiTl, the electrostatic energy (both *ionic* and *metallic*) favors the CsCl-type structure and the *covalent* interaction favors the NaTl-type structure. From LiAl to LiTl, the *covalent* interaction gets increasingly weaker. In LiAl, *covalency* overrules electrostatic energy and the NaTl-type structure is preferred. In LiTl, *ionicity* and *metallicity* dictate the structure and the CsCl type is adopted.

4.2 Alkali Metal Thallides

Among LiTl, NaTl, and KTl, only NaTl matches the classical *pseudoatom* rationalization. LiTl does not follow the 8-*N* rule, which, as discussed in last section, occurs because *covalency* loses the competition to *ionicity* and *metallicity*. The structure of KTl [161] (Fig. 17) bears no resemblance to either LiTl (CsCl type) or NaTl. It has an orthorhombic structure and Tl atoms form distorted octahedra separated by K atoms. The formal $[Tl_6]^{6-}$ distorted octahedral clusters, local point symmetry C_{2v}, adopt a distorted cubic closest packing arrangement with K atoms in tetrahedral holes. The DOS calculated with LMTO (Fig. 20) indicates that KTl should be metallic, with the Fermi level located at the bonding–antibonding crossover of the COHP curve for near neighbor Tl–Tl interactions, indicating optimization of covalent bonding within the electronegative component.

In an octahedron, each Tl atom can still be viewed as connected to four other Tl atoms, so the octet rule remains formally valid. Then, why KTl does not adopt the NaTl-type structure? Actually, it does, but not under ambient conditions. Evers and Oehlinger [179] observed NaTl-type KTl at pressures higher than 2 kbar. To investigate the effect of pressure, we calculated energy vs. volume curves for *A*Tl (*A* = Li, Na, K) in the three different structure types using VASP (Fig. 20). Above all, VASP predicts the right structures for every case. Moreover, the *E*(*V*) curve of KTl shows that, indeed, it can transform into the NaTl-type structure under pressure; the predicted pressure at 0 K is 1.027 GPa (10.27 kbar), which is five times higher than the experimental value (2 kbar) required at room temperature [179]. Furthermore, the *E*(*V*) curve of KTl reveals that a CsCl-type structure will

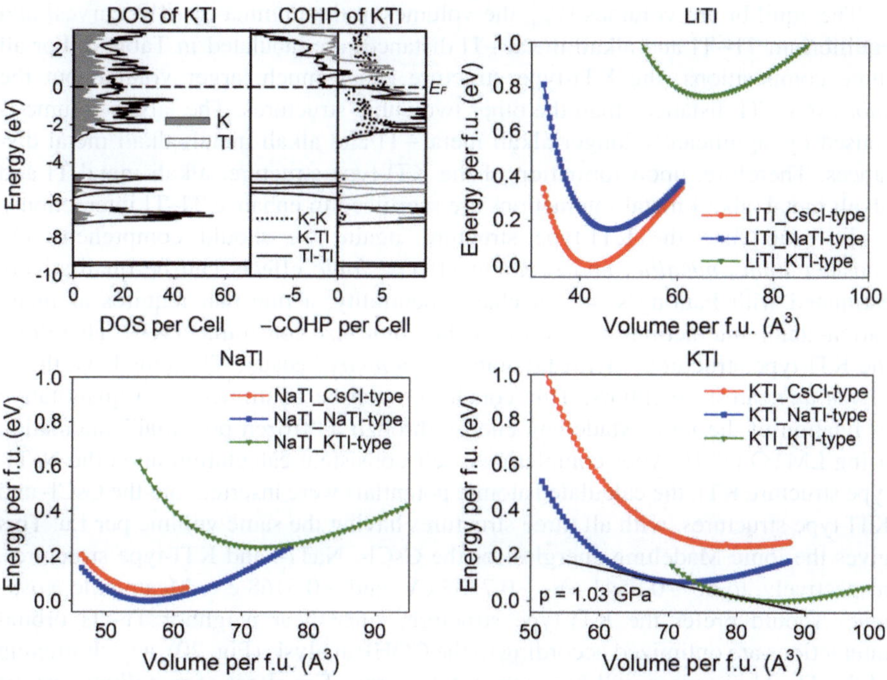

Fig. 20 DOS and COHP curves of KTl and E vs. V curves of alkaline thallides

Table 4 Equilibrium volumes per f.u. and equilibrium interatomic distances obtained from $E(V)$ curves for LiTl [78], NaTl [19], and KTl [161]

		CsCl type		NaTl type		KTl type	
		VASP	Exp.	VASP	Exp.	VASP	Exp.
LiTl	V_{eq} (Å^3)	42.66	40.53	45.92		65.79	
	r_{Tl-Tl} (Å)	3.494	3.453	3.101		2.883–3.270	
	r_{Li-Tl} (Å)	3.026	2.990	3.101		3.408–3.555	
	r_{Li-Li} (Å)	3.494	3.453	3.101		3.822–4.092	
NaTl	V_{eq} (Å^3)	57.29		56.26	51.61	72.83	
	r_{Tl-Tl} (Å)	3.855		3.318	3.224	2.982–3.383	
	r_{Na-Tl} (Å)	3.339		3.318	3.224	3.526–3.678	
	r_{Na-Na} (Å)	3.855		3.318	3.224	3.954–4.233	
KTl	V_{eq} (Å^3)	80.00		71.74		85.36	78.32
	r_{Tl-Tl} (Å)	4.309		3.599		3.144–3.567	3.055(2)–3.466(3)
	r_{K-Tl} (Å)	3.732		3.599		3.717–3.878	3.612(7)–3.768(4)
	r_{K-K} (Å)	4.309		3.599		4.169–4.463	4.051(8)–4.337(6)

not be obtained even under pressure – its energy is always higher than the NaTl-type structure. This matches Evers and Oehlinger's work, too: the CsCl-type structure was not found up to 45 kbar [179].

The equilibrium volumes [V_{eq}, the volumes at the minima of $E(V)$ curves] and equilibrium Tl−Tl and alkali metal−Tl distances are tabulated in Table 4. For all three compositions, the KTl-type structure has a much larger volume but the *shortest* Tl–Tl distances than the other two cubic structures. The larger volume is caused by significantly longer alkali metal−Tl and alkali metal–alkali metal distances. Therefore, upon formation of the KTl-type structure, alkali metal-Tl and alkali metal–alkali metal interactions are sacrificed to enhance Tl–Tl interactions.

To rationalize the KTl-type structure, again, we should comprehensively consider *ionic*, *metallic*, and *covalent* effects. *Ionic* effects can be qualitatively estimated with Pauling's rule of charge neutrality, a rule that requires as many cation–anion interactions as possible within ionic size constraints [180]. Therefore, the KTl-type structure is least favorable by *ionicity* because Tl atoms have themselves as nearest neighbors. This conclusion can be confirmed by a quantitative evaluation of the ionic Madelung energy through a "frozen potential" calculation using LMTO [177]. After completing a self-consistent calculation upon the NaTl-type structure KTl, the calculated atomic potentials were inserted into the CsCl- and KTl-type structures, with all three structures having the same volume per f.u. This gives the ionic Madelung energies for the CsCl-, NaTl-, and KTl-type structures, respectively, to be −0.8541 eV, −0.7333 eV, and −0.5168 eV. Meanwhile, *covalency* should prefer the KTl-type structure: since near neighbor Tl−Tl orbital interactions are optimized according to the COHP analysis (Fig. 20), any shortening of the Tl−Tl distances will lower the total energy. The effect of *metallicity* can be evaluated through the metallic electrostatic energy calculated with VASP (Table 5), values which show that the KTl-type structure is the least favored for all three compositions, as in the ionic situation. So, again, the isotropic *metallicity* and *ionicity* work in the same direction – they both prefer the two cubic structures over the KTl-type structure. VASP also shows that the electronic energy favors the KTl-type structure, which confirmed the preference of the KTl-type structure by *covalency*.

Table 5 The Madelung, electronic, and total energies of the CsCl-, the NaTl-, and the KTl-type structures calculated with VASP for alkali metal thallides (the energies of the CsCl-type structure are taken as reference). The volumes per f.u. of model structures are obtained from the energy minima of $E(V)$ curves

		CsCl type	NaTl type	KTl type
LiTl	E_{ES} (eV)	0	+64.0594	+354.6428
	$E_{electronic}$ (eV)	0	−63.8987	−353.8829
	E_{total} (eV)	0	+0.1607	+0.7599
NaTl	E_{ES} (eV)	0	+33.6416	+288.3062
	$E_{electronic}$ (eV)	0	−33.6812	−288.1067
	E_{total} (eV)	0	−0.0396	+0.1995
KTl	E_{ES} (eV)	0	−29.6066	+159.3199
	$E_{electronic}$ (eV)	0	+29.4482	−159.5624
	E_{total} (eV)	0	−0.1584	−0.2425

Once more, we can see the competition between *ionicity*, *metallicity*, and *covalency*. Among these three structure types, the CsCl type is the favorite for *ionicity* and *metallicity*; the KTl type is the favorite for *covalency*; and the NaTl type is not an optimum for either *ionicity/metallicity* or *covalency*, but, rather, is the best compromise among all three. The shift in dominance from *isotropic* ionicity/metallicity to *anisotropic* covalency can be understood, in part, by the size effect of the alkali metals. From LiTl to KTl, due to the increase in the atomic radius of the alkali metal, the unit cells for all three structures get larger, an effect which increases the average interatomic distances [r_0 in (1) and R_a in (2)] and raises the electrostatic energy, becoming less negative. Hence, the electrostatic energy loses its proportional contribution to the total energy from LiTl to KTl. At the same time, by switching from the cubic to the KTl-type structure, there is an additional energy penalty for the Madelung term. The overall penalty gets smaller from LiTl to KTl: in LiTl, 354.6 eV per f.u.; in NaTl, 288.3 eV per f.u.; and in KTl, 159.3 eV per f.u. On the other hand, as mentioned above, the orthorhombic KTl-type structure affords shorter Tl–Tl distances than the two cubic structures (Table 4), and any shortening of the Tl–Tl distance will lower the total energy because Tl–Tl covalent interactions are optimized. Such shortening of the Tl–Tl distances gets more significant from LiTl to KTl: in LiTl, d_{Tl-Tl}(NaTl type) – d_{Tl-Tl}(KTl type) = 0.219 Å; in NaTl, 0.336 Å; and in KTl, 0.455 Å. Therefore, along the sequence from the cubic CsCl type and NaTl type to the KTl-type structures, the energy benefit through *covalency* enhancement increases from LiTl to KTl.

This explains why the structural determining term switches from the *isotropic* ionicity/metallicity to *anisotropic* covalency and the structure transforms from cubic to the KTl-type when the alkali metal gets larger. It can also explain the NaTl-type KTl under high pressure. By applying pressure, average interatomic distances decrease and the Madelung energy drops, i.e., becomes more negative. The proportion of the Madelung term to the total energy increases and regains its structural dominance.

There is another important factor that should be considered for the electronic structure and energetic analysis of thallides. Besides the scalar contraction of the valence *s* orbital, relativistic effects also enhance spin–orbit coupling for *p*, *d*, and *f* orbitals especially in heavy atoms like Tl, an effect which can be essential to understand the structures and properties of compounds containing such 6th period elements. A recent work by Jansen et al. [181] studied $Cs_{18}Tl_8O_6$, which features tetrahedral star $[Tl_8]^{6-}$ clusters. Without consideration of spin–orbit coupling, the computed electronic structure yielded an open-shell configuration, which contradicted its experimentally observed diamagnetism. But, when spin–orbit coupling was included in the calculation, a close-shell configuration was obtained. Moreover, calculations on an isolated tetrahedral star $[Tl_8]^{6-}$ cluster revealed that spin–orbit coupling lowers the total electronic energy by 37 kJ/mol.

To evaluate the effect of spin–orbit coupling in alkali metal thallides, we compare their DOS curves and total energies obtained from LMTO calculations with and without spin–orbit coupling in Fig. 21. Spin–orbit coupling renders deeper pseudogaps at the corresponding Fermi levels in the DOS curves of NaTl and KTl, but does not significantly affect the DOS of LiTl. Furthermore, spin–orbit coupling lowers the total energy about 30 meV per f.u. more in NaTl and KTl than in LiTl. These results suggest an decreasing impact of spin–orbit coupling effects on the electronic structure and total energy of heavy atom systems under compression, either by chemical means as in LiTl vis-à-vis NaTl or KTl or by hydrostatic pressure. This suggestion awaits further verification by experimental and theoretical work.

4.3 Li_2AuBi and Na_2AuBi

Alkali metal gold bismuthides are isoelectronic with alkal metal thallides. Li_2AuBi [182] adopts a cubic structure, which is a derivative of the NaTl-type structure

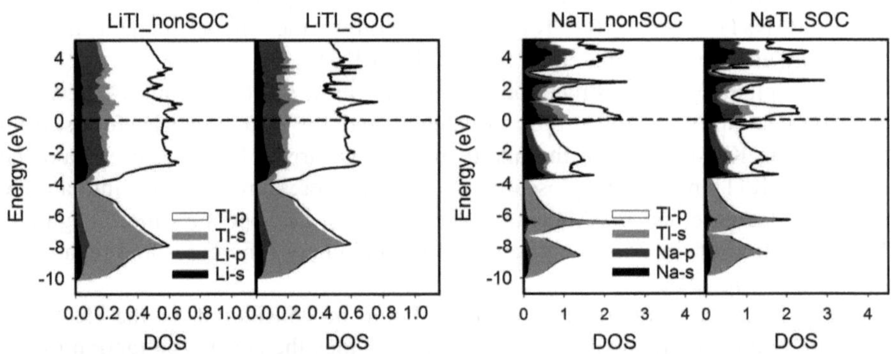

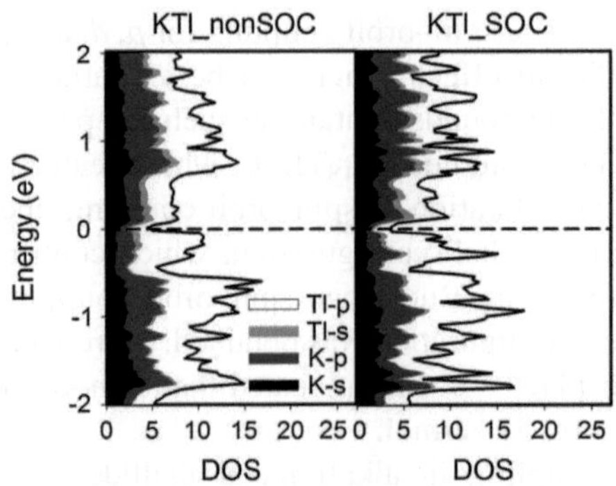

Compound	$E_{TOT,SOC} - E_{TOT,nonSOC}$ (eV/f.u.)
LiTl	−0.1519
NaTl	−0.1829
KTl	−0.1809

Fig. 21 DOS curves and total energies calculated with and without spin-orbit coupling for alkali metal thallides

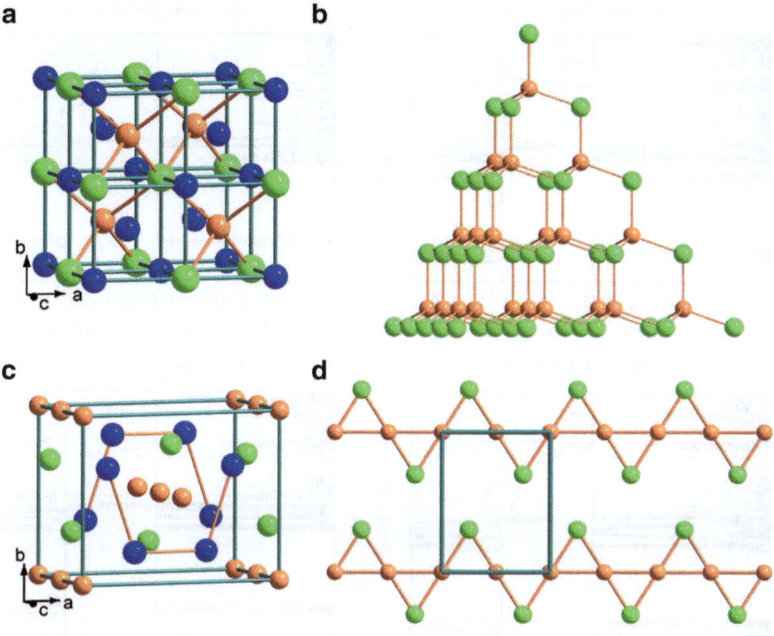

Fig. 22 Crystal structures of Li_2AuBi and Na_2AuBi: (a) the unit cell of Li_2AuBi; (b) the diamond framework of Au-Bi in Li_2AuBi; (c) the unit cell of Na_2AuBi; and (d) the 1-D "ribbon" composed of Au-Bi (black/blue: Li and Na; medium gray/golden: Au; and light gray/green: Bi)

(Fig. 22a, b), in which Au and Bi atoms form a diamond network with only heteroatomic (Au–Bi) contacts. The Zintl–Klemm formalism works perfectly for this structure. Assuming Li donates its valence electron, the average valence electron concentration (*vec*) of Au and Bi is 4, so, on average, every Au or Bi atom forms covalent bonds with 4 neighbors. If Au is also considered as a one-electron donor, i.e., Au(I) with $(5d)^{10}$ valence configuration, the *vec* of Bi atom is 8, which implies no homoatomic Bi–Bi bonds. By substituting Li for Na or K, a significantly different, orthorhombic structure emerged (Fig. 22c, d) [183]. Here Au and Bi atoms form 1D zigzag ribbons, to which the octet rule does not apply. These ribbons consist of linear chains of Au atoms, bridged by Bi atoms on alternating sides of the chain; such structures are observed in AuCl [184], AuBr [185], and AuI [186].

The electronic structures of these two compounds are shown with the DOS and COHP curves in Fig. 23. Above all, by contrast with the low-lying semicore Tl 5*d* states in NaTl (Fig. 17), we can see that the 5*d* states of Au interact significantly with Bi 6*p* states. This is in accordance with the periodicity of the elements, as demonstrated by calculating the energy band centers of 6*s*, 6*p*, and 5*d* orbitals for three compositions, Na_2AuBi, "Na_2HgPb," and Na_2Tl_2, in three structure types, CsCl-type, NaTl-type, and orthorhombic Na_2AuBi-type. These results, illustrated in Fig. 24, show that in Na_2Tl_2, the 6*s*, 6*p*, and 5*d* states are well separated from each other; in "Na_2HgPb," the 5*d* states from Hg are very close to Pb 6*s* orbitals; for

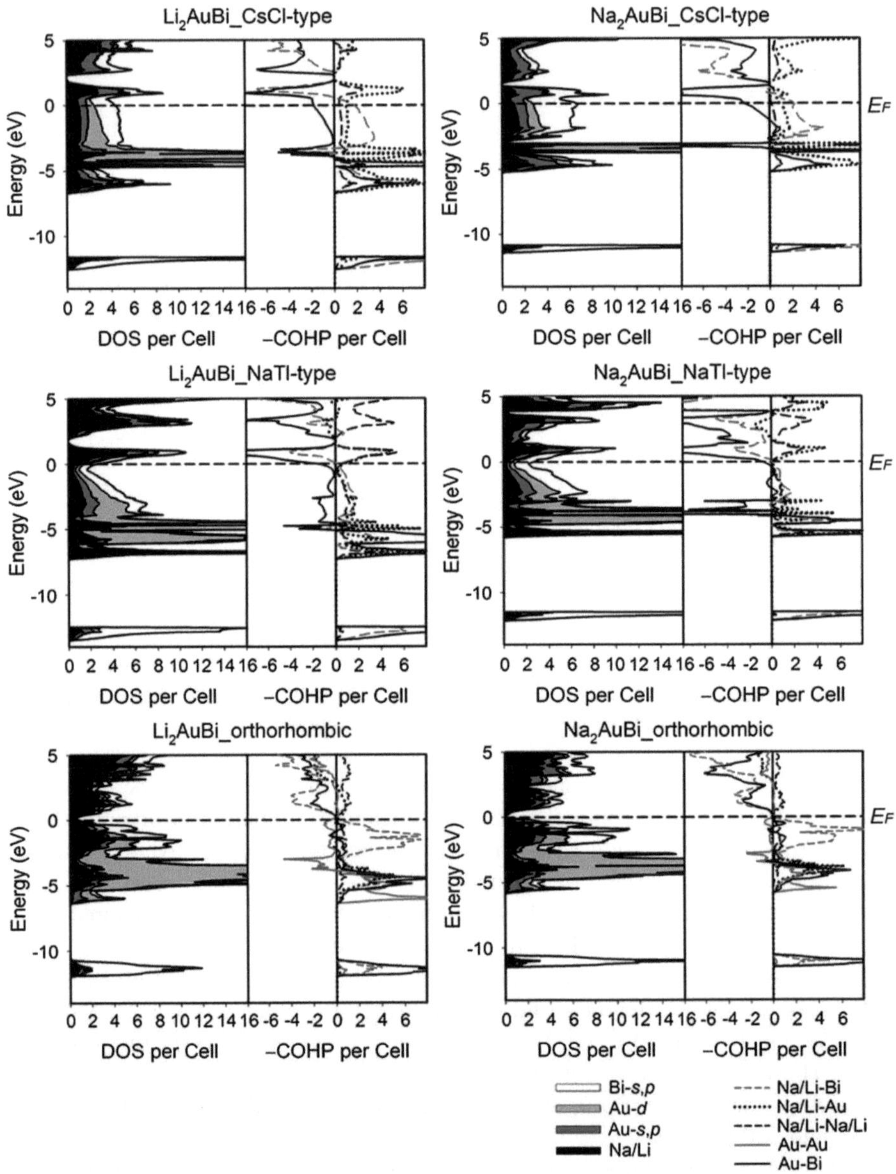

Fig. 23 DOS and COHP curves of Li_2AuBi and Na_2AuBi in three structure types

Na_2AuBi, the 5*d* states of Au are close to the 6*p* states of Bi. Thus, in Tl–Tl pairs 6*p*–6*p* interactions dominate covalent interactions and shift toward enhanced 5*d*–6*s* interactions within Hg–Pb pairs, on to 5*d*–6*p* interactions in Au–Bi pairs. Such active involvement of the 5*d* states in Au explains why the octet rule does not work well in Na_2AuBi – this rule applies when only *s* and *p* valence states are included. Violating

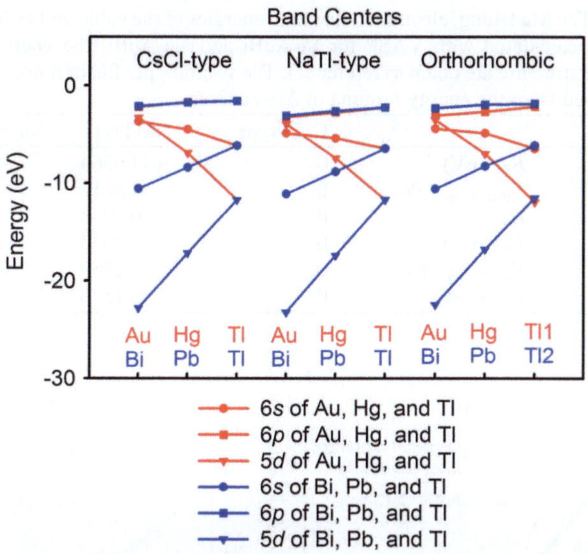

Fig. 24 The occupancy weighted band centers of Na_2AuBi, Na_2HgPb, and NaTl in the three structure types (calculated with LMTO)

Table 6 Equilibrium volumes per f.u. and interatomic distances in Li_2AuBi [181] and Na_2AuBi [182] obtained from $E(V)$ curves calculated by VASP

		CsCl type	NaTl type		Na_2AuBi type	
		VASP	VASP	Exp.	VASP	Exp.
Li_2AuBi	V_{eq} (Å^3)	74.26	79.40	73.52	86.73	–
	$r_{Au–Bi}$ (Å)	3.336	2.954	2.880	2.681	–
	$r_{Li–Li}$ (Å)				3.349	–
	$r_{Li–Au}$ (Å)	2.889			3.322, 3.489	–
	$r_{Li–Bi}$ (Å)				3.098–3.311	–
	$r_{Au–Au}$ (Å)	4.718	–	–	2.848	–
Na_2AuBi	V_{eq} (Å^3)	96.78	94.65	–	111.67	106.35
	$r_{Au–Bi}$ (Å)	3.644	3.133	–	2.797	2.752
	$r_{Na–Na}$ (Å)			–	3.494	3.438
	$r_{Na–Au}$ (Å)	3.156		–	3.466, 3.641	3.412, 3.582
	$r_{Na–Bi}$ (Å)			–	3.232–3.455	3.180–3.400
	$r_{Au–Au}$ (Å)	5.153	–	–	2.972	2.924

the octet rule, however, does not mean an absence of *covalent* interactions in Na_2AuBi. Actually, the COHP curve of Au–Bi clearly shows optimization of covalent bonding. Na_2AuBi emphasizes that the essence of the Zintl–Klemm concept need not be the *octet rule*, but, rather, the presence of optimized *covalent* interactions. The DOS and COHP curves also show that the "cations" are not just electron donors and structural "spectators": Li and Na have noteworthy contributions to the states below the Fermi level, and Li–Bi and Na–Bi bonding interactions are optimized.

The structural differences between Li_2AuBi and Na_2AuBi really trigger curiosity. An analogy can be made with the difference between LiTl, NaTl, and KTl.

Table 7 The Madelung, electronic, and total energies of the cubic and orthorhombic structures calculated with VASP for Li_2AuBi and Na_2AuBi (the energies of the CsCl-type structure are taken as reference). The volumes per f.u. of model structures are obtained from the energy minima of *E*(*V*) curves

		CsCl type	NaTl type	Na_2AuBi type
Li_2AuBi	E_{ES} (eV)	0	+41.6564	+201.1903
	$E_{electronic}$ (eV)	0	−41.6612	−200.8118
	E_{total} (eV)	0	−0.0048	+0.3785
Na_2AuBi	E_{ES} (eV)	0	+66.0933	+150.4878
	$E_{electronic}$ (eV)	0	−66.2459	−151.2092
	E_{total} (eV)	0	−0.1526	−0.7214

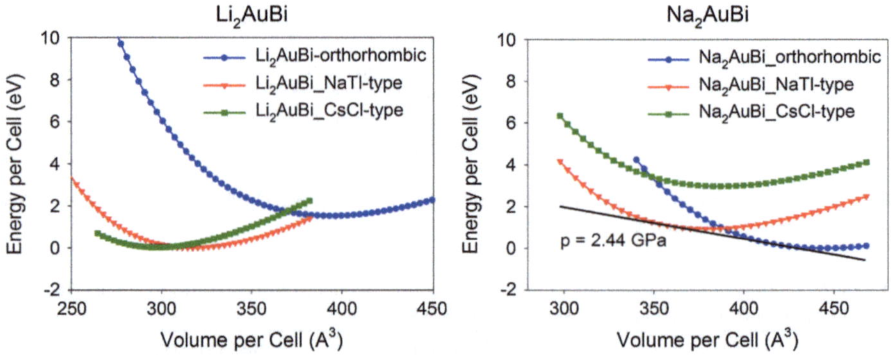

Fig. 25 *E*(*V*) curves of Li_2AuBi and Na_2AuBi

In both cases, by increasing the size of alkali metal, the 3D network transforms into lower dimensional structures. More analogies were found after we calculated the *E*(*V*) curves (Fig. 25) with VASP. The orthorhombic structure has a much larger V_{eq} but a much shorter Au–Bi distance than the two cubic structures (Table 6). The *E*(*V*) curves also predict that, just like KTl, Na_2AuBi may transform into the NaTl-type structure under an elevated pressure estimated at 2.44 GPa. By partitioning the total energy (Table 7), the two cubic structures are favored by the Madelung (metallic) term, and the orthorhombic structure is favored by the electronic term. According to Pauling's rule of charge neutrality [180], the ionic interaction also favors the two cubic structures. Therefore, the difference between Li_2AuBi and Na_2AuBi occurs because the structural dominance switches from *ionicity* and *metallicity* toward *covalency*.

Generalizing the discussions above, we can see that, indeed, to thoroughly rationalize the structures of Zintl phases with the Zintl–Klemm concept, we need to evaluate comprehensively the effects of *metallicity* and *ionicity* besides *covalency*. For all examples above, *metallicity* and *ionicity* prefer the structure with higher symmetry and 3D networks, while *covalency* counteracts with them and prefers lower symmetry and lower dimensional structural motifs. By tuning electronegativity, atomic size, or pressure, one may change the relative contributions of the Madelung and electronic terms in the total energy expression and achieve structure variation. It is not necessary that all Zintl phases follow the octet rule, for instance, when *d* states are involved in bonding.

5 Future Outlook

Why should Zintl phases continue to receive attention? Many recent investigations into new materials and inorganic solids typically engender the exploitation of specific properties, e.g., superconducting, thermoelectric, or magnetoresponsive behavior. Many of these characteristics depend on the DOS value, $N(E_F)$ or DOS slope, $|dN/dE|_{E_F}$ at the Fermi level. As a general classification of solids, Zintl phases have not been exclusively identified with many important physical properties, although they have recently demonstrated ability to form unusual polyanionic hydrides [187, 188]. Of course, such phases allow continued investigation into the intricacies of chemical bonding, nuances of which are vital for the valid application of the Zintl–Klemm formalism. In particular, as illustrated by the DOS curves in the previous sections of this chapter, Zintl phases give Fermi levels that fall within pseudogaps. Slight changes in valence electron counts, which may occur by chemical substitutions, can shift the Fermi level and lead to interesting physical properties because of increasing values of either $N(E_F)$ or $|dN/dE|_{E_F}$. Of course, subtle changes in orbital interactions, as indicated by a COHP analysis of the DOS, can also take place with such substitutions.

Table 8 Summary of existing *AX* (*A* = alkali metal; *X* = dodecalides, trielide, or tetrelide element)

	Li	Na	K	Rb	Cs
Zn	NaTl type (*cF*16)				
Cd	NaTl type (*cF*16)				
Hg	CsCl type (*cP*2)	NaHg type (*oC*16)	KHg type (*aP*8)		KHg type (*aP*8)
Al	NaTl type (*cF*16)				
Ga	NaTl type (*cF*16); $I4_1/amd$ (*tI*8)				
In	NaTl type (*cF*16); $I4_1/amd$ (*tI*8)				
Tl	CsCl type (*cP*2)	NaTl type (*cF*16)	KTl type (*oP*48); NaTl type (*cF*16)		CsTl type (*oP*96)
Si	MgGa type (*tI*16)	NaSi type (*mC*8)	KGe type (*cP*16)	KGe type (*cP*16)	KGe type (*cP*16)
Ge	MgGa type (*tI*16); $I4_1/amd$ (*tI*24)	$P12_1/c1$ (*mP*32)	KGe type (*cP*16)	KGe type (*cP*16)	KGe type (*cP*16)
Sn	LiSn type (*mP*6); $I4_1/amd$ (*tI*24)	NaPb type (*tI*16)	NaPb type (*tI*16)	NaPb type (*tI*16)	NaPb type (*tI*16)
Pb	CsCl type (*cP*2); $R\bar{3}m$(*hP*2)	NaPb type (*tI*16)	NaPb type (*tI*16)	NaPb type (*tI*16)	

Table 9 The crystal structures of existing *AX* (*A* = alkali metal; *X* = dodecalide, trielide, or tetrelide element)

Compound	Structure type	Structure	Memo
NaHg	NaHg type (*oC*16)	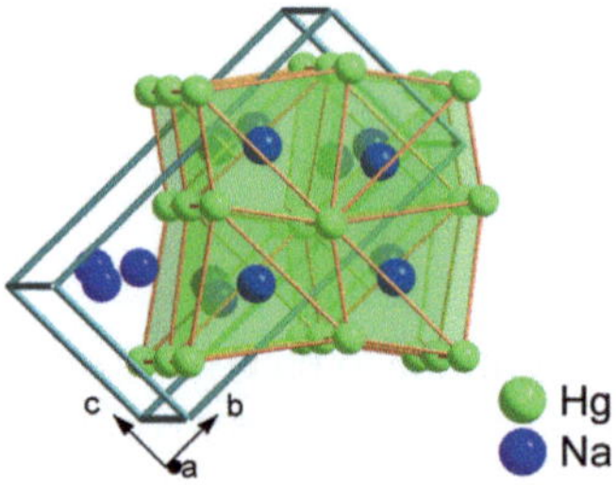	Distorted CsCl type structure [189]
KHg	KHg type (*aP*8)	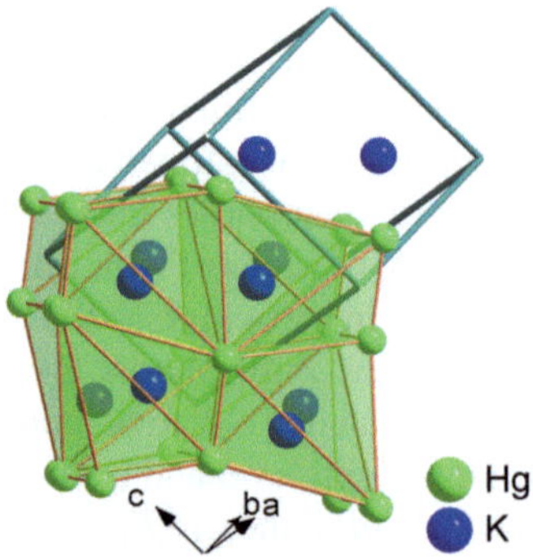	Distorted CsCl-type structure [190]
LiAl (LT)	$I4_1/amd$ (*tI*8)	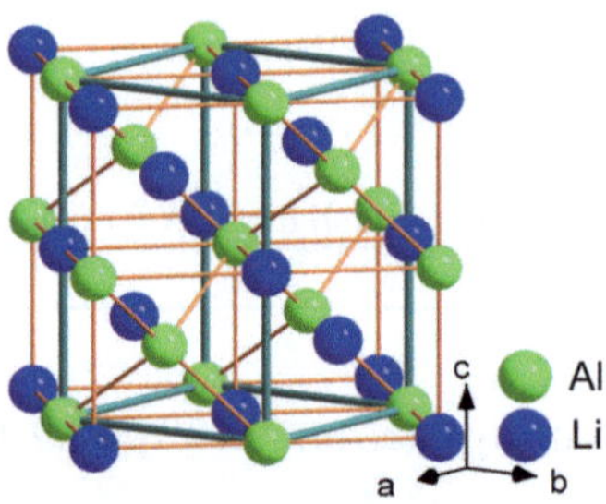	Distorted NaTl-type structure, $c/(\sqrt{2}a) = 1.0017$; form at low temperature [79, 191]
LiSi	MgGa type (*tI*32)	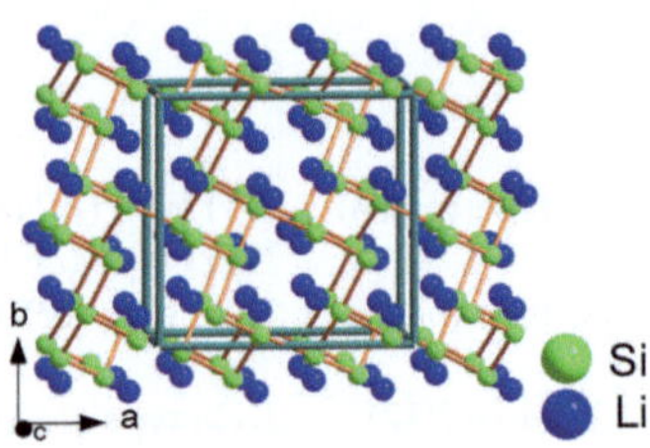	3-bonded Si atoms [192]

(*continued*)

Table 9 (continued)

Compound	Structure type	Structure	Memo
LiGe (HP)	$I4_1/amd$ ($tI24$)	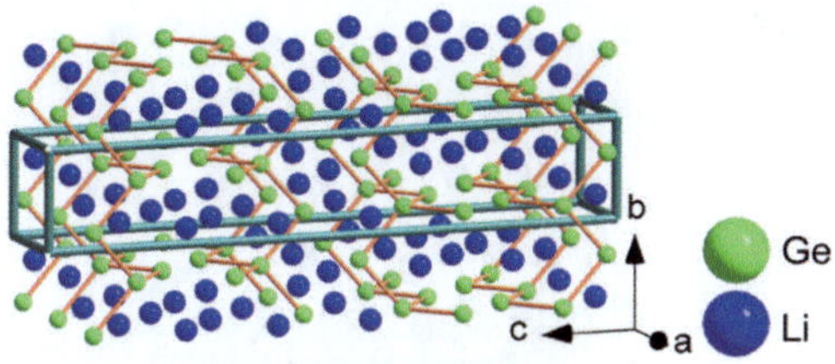	High pressure phase [193, 194]
LiSn	LiSn type ($mP6$)	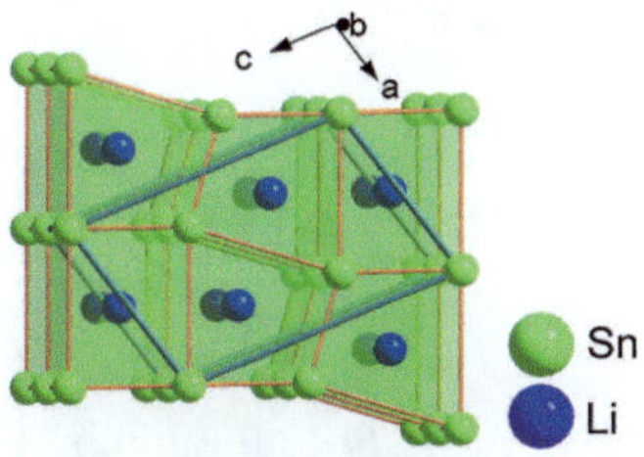	Distorted CsCl-type structure [195]
LiPb	$R\bar{3}m$ ($hP2$)	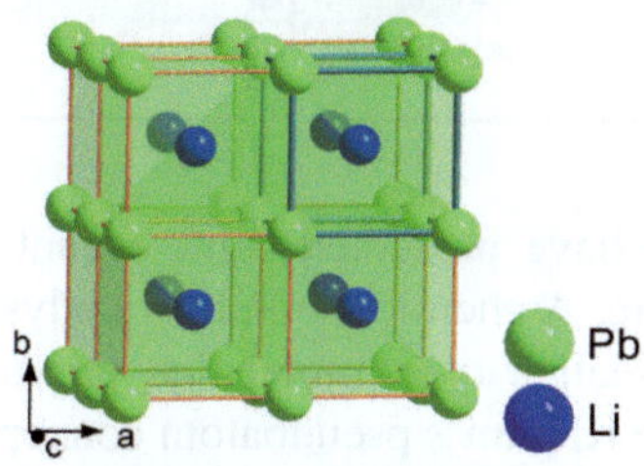	Distorted CsCl-type structure; $\alpha = 89.50(5)°$ [196]
NaSi	NaSi type ($mC32$)	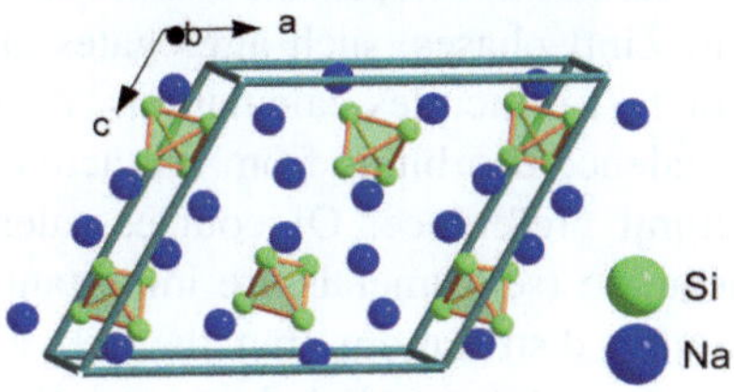	Si_4 tetrahedra [197]
NaGe	$P12_1/c1$ ($mP32$)	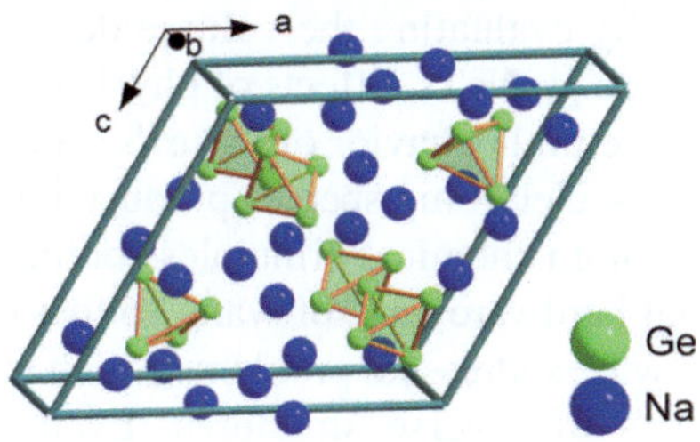	Si_4 tetrahedra [198]

(*continued*)

Table 9 (continued)

Compound	Structure type	Structure	Memo
NaSn	NaPb type ($tI64$)	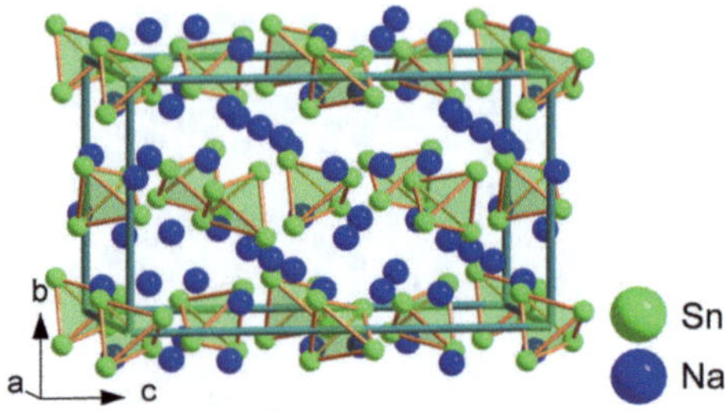	Si_4 tetrahedra [198]
KSi	KGe type ($cP64$)	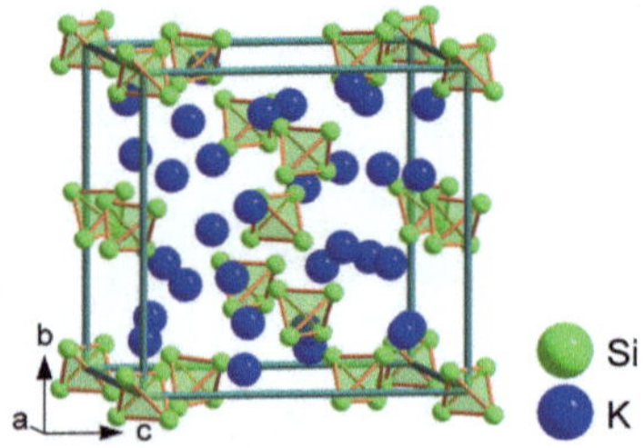	Si_4 tetrahedra [199]

In this chapter, we have presented some recent quantitative advances of the Zintl–Klemm formalism. A chemical bonding analysis of ab initio quantum chemical calculations on realgar-type molecules provides a basis set independent approach to investigate Klemm's pseudoatom concept. Although applied to neutral molecules in this case, further developments can occur by treating formally charged molecular clusters; in Zintl phases, such aggregates are typically anionic, which creates challenges for first-principles calculations. A second emphasis illustrated the influence of the valence *d* orbitals from the active metal component in Zintl phases toward structural preference. Of course, valence electron counts at the network of electronegative (semi)metals are important, especially for controlling compositions and atomic distributions, but the active metals provide additional forces that must be accurately included in any theoretical assessment. Such forces were discussed by evaluating the volume dependences of total energies of traditional Zintl phases, e.g., NaTl. Effects of physical or chemical pressure significantly influence the structural behavior of these "simple" structures; even chemical substitutions to create isoelectronic species produce unusual effects.

Applications of quantum chemical principles continue to grow through development of computational hardware and software, so that increasingly complex structural problems can now be addressed, problems which include bonding features and physical characteristics of diverse structures. Even compounds with "simple" chemical compositions, e.g., *AX*, manifest tremendous diversity, examples which are listed in Table 8 and illustrated in Table 9. Among this summary, we note that several established structures are revealing other "colors," for example, the low-temperature distortion of LiIn to a tetragonal structure [79, 191]. Thus, as long as experimental techniques also continue to develop by providing improved sensitivity to structure and physical properties, Zintl phases will show interesting properties, some of which may achieve useful technological applications.

Acknowledgments This work was supported by NSF DMR 02-41092 and 06-05949. MWS was supported by a DOE Chemical Physics program, a NSF Petascale Applications grant, and a NSF Cyberinfrastructure grant. Programming for this method was done during a visit to Prof. Keiko Takano's group at Ochanomizu University in Tokyo.

References

1. Myers R (2003) The basics of chemistry. Greenwood, Westport, CT, pp 74–77
2. Mitchell PR, Parish RV (1969) J Chem Educ 46:811
3. Gillespie RJ, Hargittai I (1991) The VSEPR model of molecular geometry. Allyn and Bacon, Boston
4. Gillespie RJ, Robinson EA (2005) Chem Soc Rev 34:396
5. Wade K (1976) Adv Inorg Chem Radiochem 18:1
6. Mingos DMP (1977) Adv Organomet Chem 15:1
7. Burdett JK (1980) Molecular shapes. Wiley, New York
8. Schäfer H, Eisenmann B, Müller W (1973) Angew Chem Intl Ed Engl 12:694
9. Schäfer H, Eisenmann B (1981) Rev Inorg Chem 3:29
10. Schäfer H (1985) Annu Rev Mater Sci 15:1
11. Nesper R (1990) Prog Solid State Chem 20:1
12. Zintl E, Dullenkopf W (1932) Z phys Chem B16:183
13. Zintl E, Brauer G (1933) Z phys Chem B20:245
14. Zintl E (1939) Angew Chem 52:1
15. Hume-Rothery W (1926) J Inst Metals 35:296
16. Westgren AF, Phragmén G (1926) Z Metallkunde 18:279
17. Laves F, Witte H (1936) Metallwirtschaft 15:840
18. Grimm HO, Sommerfeld A (1926) Z Phys 36:36
19. Zintl E, Woltersdorf G (1935) Z Elektrochem 41:876
20. Pauling L (1960) The nature of the chemical bond, 3rd edn. Cornell University Press, Ithaca, New York
21. Klemm W (1959) Proc Chem Soc London 329
22. Klemm W (1956) Trab Reun Int React Solidos 3rd 1:447
23. von Schnering HG (1985) Nova Acta Leopold 59:165
24. Kauzlarich SM (ed) (1996) Chemistry, structure, and bonding of zintl phases and ions. VCH, Weinheim, Germany
25. Miller GJ, Lee C-S, Choe W (2002) In: Meyer G, Naumann D, Wesemann L (eds) Inorganic chemistry highlights. Wiley, Weinheim
26. Allen LC (1989) J Am Chem Soc 111:9003
27. Mann JB, Meek TL, Allen LC (2000) J Am Chem Soc 122:2780
28. van Arkel AE (1956) Molecules and crystals in inorganic chemistry. Interscience, New York
29. Ketelaar JAA (1958) Chemical constitution: an introduction to the theory of the chemical bond, 2nd edn. Elsevier, Amsterdam
30. Burdett JK (1995) Chemical bonding in solids. Oxford University Press, Oxford
31. Pettifor D (1995) Bonding and structure of molecules and solids. Clarendon, Oxford
32. Palmer MH, Findlay RH (1983) J Mol Struct (Theochem) 104:327–339
33. Lu C-S, Donohue J (1944) J Am Chem Soc 66:818–827
34. Almond MJ, Forsyth GA, Rice DA et al (1989) Polyhedron 8:2631–2636

35. Sharma BD, Donohue J (1963) Acta Crystallogr 16:891–897
36. DeLucia ML, Coppens P (1978) Inorg Chem 17:2336–2338
37. Irsen SH, Jacobs P, Dronskowski R (2001) Z Anorg Allg Chem 627:321–325
38. Pritchina EA, Gritsan NP, Zibarev AV et al (2009) Inorg Chem 48:4075–4082
39. Chung G, Lee D (2002) J Mol Struct (Theochem) 582:85–90
40. Bragin J, Evans MV (1969) J Chem Phys 51:268–277
41. Turowski A, Appel R, Sawodny W et al (1978) J Mol Struct 48:313–323
42. Steudel R (1981) Z Naturforsch A36:850–858
43. Findlay RH, Palmer MH, Downs AJ et al (1980) Inorg Chem 19:1307–1314
44. Tang T-H, Bader RFW, MacDougall PJ (1985) Inorg Chem 24:2047–2053
45. Rzepa HS, Woollins JD (1990) Polyhedron 9:107–111
46. Schugart KA (1990) Polyhedron 9:1935–1937
47. Scherer W, Spiegler M, Pedersen B et al (2000) Chem Commun 635–636
48. Bridgeman AJ, Cavigliasso G, Ireland LR et al (2001) J Chem Soc Dalton Trans 2095–2108
49. Geier J (2006) J Phys Chem A110:9273–9281
50. Hassel O, Viervoll H (1943) Tidsskr Kjemi Bergv 3:7
51. Lippincott ER, Tobin MC (1953) J Chem Phys 21:1559–1565
52. Drew MGB, Templeton DH, Zalkin A (1967) Inorg Chem 6:1906–1910
53. Gieren VA, Dedderer B, Roesky HW et al (1978) Z Anorg Allg Chem 440:119–129
54. Konu J, Bajorek T, Laitinen RS et al (2006) Eur J Inorg Chem 2951–2958
55. Griffin AM, Minshall PC, Sheldrick GM (1976) J Chem Soc Chem Commun 809–810
56. Minshall PC, Sheldrick GM (1978) Acta Crystallogr B34:1326–1328
57. Ito T, Morimoto N, Sadanaga R (1952) Acta Crystallogr 5:775–782
58. Porter EJ, Sheldrick GM (1972) J Chem Soc Dalton Trans 13:1347–1349
59. Ohse L, Somer M, Brockner W (1988) Z Naturforsch A43:326–330
60. Ystenes M, Menzel F, Brockner W (1994) Spectrochim Acta A50:225–231
61. Banerjee A, Jensen JO, Jensen JL (2003) J Mol Struct (Theochem) 626:63–75
62. Bonazzi P, Menchetti S, Pratesi G (1995) Am Mineral 80:400–403
63. Fletcher GD, Schmidt MW, Gordon MS (1999) Adv Chem Phys 110:267–294
64. Perdew JP, Ruzsinsky A, Csonka GI et al (2009) Phys Rev Lett 103:026403
65. Kendall RA, Dunning TH, Harrison RJ (1992) J Chem Phys 96:6796–6806
66. P, S use the replacement known as aug-cc-pV(T+d)Z: Dunning TH, Peterson KA, Wilson AK (2001) J Chem Phys 114:9244–9253
67. Wilson AK, Woon DE, Peterson KA et al (1999) J Chem Phys 110:7667–7676
68. Schmidt MW, Baldridge KK, Boatz JA et al (1993) J Comput Chem 14:1347–1363
69. Lu WC, Wang CZ, Schmidt MW et al (2004) J Chem Phys 120:2629–2637; 120:2638–2651
70. Edmiston C, Ruedenberg K (1963) Rev Mod Phys 35:457–465
71. Raffenetti RC, Ruedenberg K, Janssen CL et al (1993) Theor Chim Acta 86:149–165
72. Ivanic J, Atchity GM, Ruedenberg K (2008) Theor Chem Acc 120:281–294
73. Ivanic J (2003) J Chem Phys 119:9364–9376; 119:9377–9385
74. Coulson CA, Longuet-Higgins HC (1947) Proc R Soc London A191:39–60
75. Lennard-Jones JE (1952) J Chem Phys 20:1024–1029
76. Chirgwin BH (1957) Phys Rev 107:1013–1025
77. Löwdin PO (1955) Phys Rev 97:1474–1489; 97:1490–1508; 97:1509–1520
78. Baden W, Schmidt PC, Weiss A (1979) Phys Stat Sol A 51:183–190
79. Ehrenberg H, Pauly H, Knapp M et al (2004) J Solid State Chem 177:227–230
80. Jang GE, Curelaru IM, Hentschel MP (1994) J Cryst Growth 141:399–403
81. Kuriyama K, Saito S, Iwamura K (1979) J Phys Chem Solids 40:457–461
82. Lindqvist I (1958) J Inorg Nucl Chem 6:159–160
83. Kim S-J, Fässler TF (2009) J Solid State Chem 182:778–789
84. Kim S-J, Kraus F, Fässler TF (2009) J Am Chem Soc 131:1469–1478
85. Klem MT, Corbett JD (2005) Inorg Chem 44:5990–5995
86. Mudring AV, Corbett JD (2005) Inorg Chem 44:5636–5640

87. Ponou S, Mueller N, Fässler TF et al (2005) Inorg Chem 44:7423–7430
88. Wendorff M, Röhr C (2007) Z Naturforsch B Chem Sci 62:1227–1234
89. Wendorff M, Röhr C (2007) Z Anorg Allg Chem 633:1541–1550
90. Xu Z, Guloy AM (1998) J Am Chem Soc 120:7349–7350
91. Ponou S, Kim S-J, Fässler TF (2007) Z Anorg Allg Chem 633:1568–1574
92. Ge M-H, Corbett JD (2007) Inorg Chem 46:4138–4144
93. Wendorff M, Röhr C (2008) Z Naturforsch B Chem Sci 63:1383–1394
94. Kim S-J, Ponou S, Fässler TF (2008) Inorg Chem 47:3594–3602
95. You T-S, Tobash PH, Bobev S (2010) Inorg Chem 49:1773–1783
96. Saparov B, He H, Zhang X et al (2010) Dalton Trans 39:1063–1070
97. Saparov B, Broda M, Ramanujachary KV et al (2010) Polyhedron 29:456–462
98. Massalski TB (ed) (1990) Binary alloy phase diagrams. ASM International, Materials Park, OH
99. Nagamatsu J, Nakagawa N, Muranaka T et al (2001) Nature 410:63
100. Hassel O (1924) Z Phys 25:317
101. Hofmann U, Wilm D (1936) Z Elektrochem 42:504
102. Trucano P, Chen R (1975) Nature 258:136
103. Lipson H, Stokes AR (1942) Nature 149:328
104. Choi HJ, Roundy D, Sun H et al (2002) Nature 418:758
105. Fogg AM, Chalker PR, Claridge JB et al (2003) Phys Rev B Condens Mat Mater Phys 67:245106/1–10
106. Karpinski J, Kazakov SM, Jun J et al (2004) Physica C Supercond 408–410:81–82
107. Kortus J, Dolgov OV, Kremer RK et al (2005) Phys Rev Lett 94:027002/1–4
108. Birajdar B, Wenzel T, Manfrinetti P et al (2006) J Phys Conf Ser 43:484–487
109. Fogg AM, Meldrum J, Darling GR et al (2006) J Am Chem Soc 128:10043–10053
110. Klein T, Lyard L, Marcus J et al (2006) Phys Rev B Condens Mat Mater Phys 73:224528/1–8
111. Mudgel M, Awana VPS, Kishan H et al (2007) Physica C Supercond 467:31–37
112. Karpinski J, Zhigadlo ND, Katrych S et al (2008) Phys Rev B Condens Mat Mater Phys 77:214507/1–11
113. Imai M, Sadki E-HS, Abe H et al (2003) Phys Rev B 68:064512
114. Imai M, Nishida K, Kimura T et al (2002) Appl Phys Lett 80:1019
115. Imai M, Nishida K, Kimura T et al (2002) Physica C 382:361
116. Lorenz B, Lenzi J, Cmaidalka J et al (2002) Physica C 383:191
117. Lorenz B, Cmaidalka J, Meng RL et al (2003) Phys Rev B 68:014512
118. Nakagawa T, Tokunaga M, Tamegai T (2006) Sci Technol Adv Mater 7:S108
119. Yamanaka S, Otsuki T, Ide T et al (2007) Physica C 451:19
120. Evans MJ, Wu Y, Kranak VF et al (2009) Phys Rev B 80:064514
121. Shein IR, Medvedeva NI, Ivansivskii AL (2003) J Phys Condens Matter 15:L541
122. Mazin II, Papaconstantopoulos DA (2004) Phys Rev B 69:180512(R)
123. Huang GQ, Chen LF, Liu M et al (2004) Phys Rev B 69:064509
124. Giantomassi M, Boeri L, Bachelet GB (2005) Phys Rev B 72:224512
125. Huang GQ, Liu M, Chen LF et al (2005) Physica C 423:9
126. Hoffmann RD, Pöttgen R (2001) Z Kristallogr 216:127–145
127. Burdett JK, Miller GJ (1990) Chem Mater 2:12–26
128. Villars P, Calvert LD (1997) Pearson's handbook of crystallographic data for intermetallic phases, 2nd edn. American Society for Metals, Materials Park, OH
129. Amerioun S (2003) Phase and structural stability principles of polar intermetallic AeE_4 and AeE_2 Systems (Ae = Ca, Sr, Ba; E = Al, Ga, In). Ph.D. thesis, Stockholm University
130. Johnston RL, Hoffmann R (1992) Z Anorg Allg Chem 616:105
131. Söderberg K, Boström M, Kubotac Y et al (2006) J Solid State Chem 179:2690–2697
132. Kal S, Stoyanov E, Belieres J-P et al (2008) J Solid State Chem 181:3016–3023
133. Teatum E, Gschneidner KA, Waber J (1960) Compilation of calculated data useful in predicting metallurgical behaviour of the elements in binary alloy systems. LA-2345, Los Alamos Scientific Laboratory

134. Debray D (1973) J Less Common Met 30:237
135. Markiv VYa, Belyavina NN, Zhunkovskaya TI (1984) J Less Common Met 97:5
136. Pöttgen R, Kotzyba G, Schappacher FM et al (2001) Z Anorg Allg Chem 627:1299–1304
137. Bojin MD, Hoffmann R (2003) Helv Chim Acta 86:1653, 1683
138. Nuspl G, Polborn K, Evers J et al (1996) Inorg Chem 35:6922–6932
139. Dronskowski R, Blochl P (1993) J Phys Chem 97:8617
140. Jepsen O, Burkhardt A, Andersen OK (1999) The TB-LMTO-ASA Program, version 4.7. Max-Planck-Institut für Festkörperforschung, Stuttgart, Germany
141. Lee S (1991) J Am Chem Soc 113:101
142. Gourdon O, Gout D, Miller GJ (2005) Electronic states of intermetallic compounds. In: Bassani G, Liedl G, Wyder P (eds) Encyclopedia of condensed matter physics. Elsevier, Amsterdam, Netherlands, pp 409–422
143. Haarmann F, Koch K, Grüner D et al (2009) Chem Eur J 15:1673–1684
144. Haarmann F, Prots Y (2006) Z Anorg Allg Chem 632:2135
145. Sichevych O, Ramlau R, Giedigkeit R, et al (2000) Abstracts 13th international conference on solid compounds of transition elements, Stresa, O-13
146. You T-S, Lidin S, Gourdon O et al (2009) Inorg Chem 48:6380–6390
147. You T-S, Miller GJ (2009) Inorg Chem 48:6391–6401
148. Bobev S, Bauer ED, Thompson JD et al (2004) J Solid State Chem 177:3545–3552
149. Emsley J (1998) The elements. Clarendon, Oxford
150. Bockelmann W, Jacobs H, Schuster HU (1974) Z Anorg Allg Chem 410:233
151. Pöttgen R (1996) J Alloys Comp 243:L1
152. Eisenmann B, Klein J, Somer M (1990) Angew Chem Int Ed Engl 29:87
153. Kresse G, Hafner J (1993) Phys Rev B 47:558–561
154. Kresse G, Hafner J (1994) Phys Rev B 49:14251–14269
155. Kresse G, Furthmüller J (1996) Comput Mater Sci 6:15–50
156. Kresse G, Furthmüller J (1996) Phys Rev B 54:11169–11186
157. You T-S, Zhao J-T, Pöttgen R et al (2009) J Solid State Chem 182:2430–2442
158. You T-S, Miller GJ (2007) Inorg Chem 46:8801–8811
159. Evers J, Oehlinger G, Weiss A et al (1983) J Less Common Met 90:L19
160. Zheng C, Hoffmann R (1989) Inorg Chem 28:1074
161. Dong Z, Corbett JD (1993) J Am Chem Soc 115:11299–11303
162. Jarrold MF, Ray U, Ijiri Y (1991) Z Phys D 19:337–342
163. Honea EC, Ogura A, Murray CA et al (1993) Nature 366:42
164. Zdetsis AD (2001) Phys Rev A 64:023202
165. Zdetsis AD (2005) Comp Lett 1:337–342
166. Grube VG, Schmidt A (1936) Ztschr Elektrochem 42:201–209
167. Dronskowski R (2005) Computational chemistry of solid state materials. Wiley, Weinheim
168. Hoppe R (1966) Angew Chem Int Ed Engl 5:95–106
169. Hoppe R (1970) Angew Chem Int Ed Engl 9:25–34
170. Waddington TC (1959) Adv Inorg Chem Radiochem 1:157–221
171. Ewald PP (1921) Ann Phys Leipzig 64:253–287
172. Ashcroft NW, Mermin ND (1976) Solid state physics. Thomson Learning, CA, USA
173. Hafner J (1987) From Hamiltonians to phase diagrams. Springer, New York
174. Fuchs K (1935) Proc R Soc Lond A 151:585–602
175. Kresse G, Hafner J (1994) J Phys Condens Matter 6:8245–8257
176. Pawlowska Z, Christensen NE, Satpathy S et al (1986) Phys Rev B 34:7080–7088
177. Christensen NE (1985) Phys Rev B 32:207–228
178. Kresse G, Joubert D (1999) Phys Rev B 59:1758–1775
179. Evers J, Oehlinger G (2000) Inorg Chem 39:628–629
180. Pauling L (1929) J Am Chem Soc 51:1010–1026
181. Wedig U, Saltykov V, Nuss J, et al. (2010) J Am Chem Soc 132:12458–12463
182. Pauly H, Weiss A, Witte H (1968) Z Metallkde 59:47–57

183. Kim S-J, Miller GJ, Corbett JD (2010) Z Anorg Allg Chem 636:67–73
184. Janssen EMW, Folmer JCW, Wiegers GA (1974) J Less Common Met 38:71–76
185. Janssen EMW, Wiegers GA (1978) J Less Common Met 57:58–67
186. Jagodzinski H (1959) Z Kristallogr 112:80–87
187. Lee MH, Bjorling T, Hauback BC et al (2008) Phys Rev B Condens Mat Mater Phys 78:195209/1–11
188. Evans MJ, Holland GP, Garcia-Garcia FJ et al (2008) J Am Chem Soc 130:12139–12147
189. Nielsen JW, Baenziger NC (1954) Acta Crystallogr 7:277–282
190. Duwell EJ, Baenziger NC (1955) Acta Crystallogr 8:705–710
191. Ehrenberg H, Pauly H, Hanse T et al (2002) J Solid State Chem 167:1–6
192. Evers VJ, Oehlinger G, Sextl G (1993) Angew Chem 105:1532–1534
193. Evers VJ, Oehlinger G, Sextl G et al (1987) Angew Chem 99:69–71
194. Blasé W, Cordier G (1990) Z Kristalogr 193:317–318
195. Müller W, Schäfer H (1973) Z Nat B Anorg Chem Org Chem 28:246–248
196. Zalkin A, Ramsey WJ (1957) J Phys Chem 61:1413–1415
197. Witte J, Schnering HG, Klemm W (1964) Z Anorg Allg Chem 327:260–273
198. Müller W, Volk K (1977) Z Nat B Anorg Chem Org Chem 32:709–710
199. Busmann E (1961) Z Anorg Allg Chem 313:90–106

Struct Bond (2011) 139: 57–96
DOI:10.1007/430_2010_29

Published online: 3 December 2010

High Pressure Investigations on A^IB^{III} Zintl Compounds (A^I = Li to Cs; B^{III} = Al to Tl) up to 30 GPa

Jürgen Evers

Abstract For the AIBIII compounds (AI = Li to Cs; BIII = Al to Tl) the stabilities of the NaTl- and the β-brass-type structures have been calculated by density functional theory (DFT) with the program WIEN2k. In agreement with the experimental data obtained at 1 bar LiAl, LiGa, NaIn and NaTl are stable with the NaTl-type structure, LiTl with the β-brass-type structure. Interestingly, LiIn with the NaTl-type structure is high temperature phase and LiIn with the β-brass-type structure is low temperature phase. For the 14 remaining AIBIII compounds the endothermic formation energies at 0 K and 0 bar have been estimated for the NaTl-type structure phases. In addition, the high pressure range is estimated to obtain such phases in the diamond anvil cell. For LiTl, NaTl, KTl, RbTl, and CsTl the results of high pressure investigations in the diamond anvil cell up to 30 GPa are reported. The near-neighbor-diagram for the AIBIII compounds with the NaTl-type structure has been calculated.

Keywords β-brass-type structure · DFT calculation · Diamond anvil cell · High pressure transformation · NaTl-type structure · Zintl compounds

Contents

J. Evers
Ludwig Maximilian University Munich, Butenandtstr. 5-13, D-81377 Munich, Germany
e-mail: Juergen.Evers@cup.uni-muenchen.de

1 Introduction

Zintl compounds are introduced as solids with a quite unique hybrid of chemical bonding. In the Zintl concept, it is assumed that in such compounds electrons are transferred from the electropositive to the electronegative constituent. By this electron transfer, the valence shell of the electronegative constituent is filled. As a result, covalent bonding in the anionic partial structure obeys then the (8-N) rule for normal valence compounds. NaTl is the prototype of Zintl compounds. Four additional A^IB^{III} compounds (LiAl, LiGa, LiIn, NaIn) with the NaTl-type structure were characterized by Zintl in the 1930s of the last century. Nevertheless, one A^IB^{III} compound, LiTl, crystallized in the alloy structure of β-brass. This structural concurrence interested the theoreticians, performing density-functional theory (DFT) calculations on such compounds, however, only on the six known ones. With KTl, synthesized in 1999 at high pressure (HP) of 2 GPa in a diamond anvil cell (DAC) by Evers and Oehlinger, a new access to such compounds was opened. In this study, the results are reported on DFT calculations for the whole 20 A^IB^{III} compounds (A^I = Li to Cs; B^{III} = Al to Tl), and, furthermore, the results of in situ HP experiments in a DAC up to 30 GPa presented for the thallides LiTl, NaTl, KTl, RbTl, and CsTl.

In the huge field of intermetallic phases, Zintl compounds are a class of solids formed by electropositive with electronegative constituents. Such polar compounds have definite compositions and a quite unique bonding hybrid with ionic, covalent, and metallic contributions. In the Zintl concept [1–6], it is assumed that valence electrons are transferred from the electropositive to the electronegative constituent. By this electron transfer, the valence shells of the electronegative partner are filled, thus forming covalent bonds. This covalent bonding in the anionic partial structure obeys the (8-N) rule of normal valence compounds. Hoffmann assigned this simple concept as the "single most important theoretical concept in solid-state chemistry of the 20th century" [7] "It forges a link between solid-state chemistry and organic, or main group chemistry" [7].

The compound NaTl [8] is considered as the prototype of Zintl phases [9], although the triel element thallium lies left to the Zintl boundary [10]. This boundary was introduced by Laves [11] as dividing line between triel elements aluminum to thallium and tetrel elements silicon to lead. It was assumed that tetrel elements form salt-like compounds with electropositive metals, contrary to triel elements that generate alloy-like compounds. Due to modern structural and electronic investigations on such compounds, especially by Corbett [10], this traditional distinction seems no longer appropriate.

NaTl, itself, crystallizes in the NaTl-type structure [8] (B32, Strukturbericht classification [12]). Assuming that one electron is transferred per formula unit (f.u.) from the electropositive A^I metal sodium to the more electronegative B^{III} element thallium, Tl^- ions become iso-valence-electronic with main group IV elements carbon to tin and build up a diamond-type partial structure with four homonuclear bonds, obeying the (8-N) rule. Zintl and coworkers prepared four additional A^IB^{III} compounds with the NaTl-type structure, such as NaIn [13], LiAl [1], LiGa [14], and LiIn [14]. However, in the homologous series of lithium compounds, LiTl [14] does not crystallize in the NaTl, but in the β-brass-type structure (B2, Strukturbericht classification [15]). As it will be shown later, the NaTl- and the β-brass-type alloy are closely related to each other, because they can be assigned as different binary varieties of a body-centered cubic (bcc) arrangement. They consist of interpenetrating cubes giving 8:8 nearest neighbor coordination for both.

The structural change from the NaTl- (B32) to the β-brass-type structure (B2), which is found when in the homologous A^ITl series the metallic radius of the A^I metal is lowered from sodium to lithium, interested a lot of theoreticians. Magnovetskii and Krasko [16] calculated with Brillouin–Wigner perturbation theory a stabilization for LiAl and NaTl in the B32 and for LiTl in the B2 structure. Schmidt [9, 17] used the local density functional approach on the basis of the muffin tin model, to calculate the structural stability of the B32 and the B2 structures, however, only for the two compounds NaTl and LiTl. Christensen [18] calculated within the density-functional formalism with the local density approximation (LDA) the stability of B2 and B32 structures in A^IB^{III} compounds LiAl, LiGa, LiIn, LiTl, NaIn, and NaTl. Christensen confirmed by such calculations the stabilities at zero temperature and pressure [18], which Zintl had found experimentally in his classical works at room temperature and 1 bar. According to Christensen, LiAl, LiGa, LiIn, NaIn, and NaTl are stable in the B32, and LiTl in the B2 structure. However, for LiIn Christensen calculated that the structural B32–B2 difference is only –2 mRy (–0.6 kcal/mole). The results of Schmidt [9, 17] and Christensen [18] agreed in the relative stabilities for the structures of NaTl and LiTl, but differed slightly in their absolute energies.

In 2002, Ehrenberg et al. [19] prepared a low temperature modification of LiIn. At 170 K cubic LiIn with the NaTl-type structure, space group Fd-3m, was transformed in a reversible group–subgroup "translationengleich" transition into a tetragonal modification in space group $I4_1/amd$. The tetragonal distortion is with 0.5% very small, however, leading to a slightly more densely packing than in the cubic case. The first structural change at high pressure (HP) from the NaTl- into the β-brass-type structure was investigated in 1998 by Schwarz et al. [20]. Starting with LiIn with the NaTl-type structure, at 11 GPa the β-brass alloy structure was obtained. Together with LiTl, stable at ambient pressure, LiIn is the second A^IB^{III} phase with the β-brass-type structure. More than 60 years after Zintl's classical works, Evers and Oehlinger prepared in 1999 with high pressure HP-KTl the first novel binary A^IB^{III} compound with the NaTl-type structure [21].

This compound was prepared at 2 GPa in a DAC from the normal pressure phase KTl which consists of isolated Tl_6-octahedra [22].

The earlier published calculations had focused the interest to only six A^IB^{III} compounds, which were firstly synthesized by Zintl. However, combining five A^I metals (Li to Cs) with four B^{III} elements (Al to Tl) results in 20 A^IB^{III} compounds. Therefore, Zintl's work, although complete at ambient pressure, is enlightening only for 30% of the whole A^IB^{III} series. However, in recent years with DACs, heated by a laser, preparative conditions are enlarged to 50 GPa and 1,000°C for routine chemical synthesizes in solid-state chemistry. Indeed, there is now a chance to have a look on the 70% of the remaining A^IB^{III} compounds for the stability of the NaTl- or the β-brass-type structure or new structures. After the successful HP-synthesis of KTl, it seems that RbTl and CsTl of the thallide series could be also good candidates for stabilizing here the NaTl-type structure after a HP treatment.

In this chapter, we report on the total energy calculations for all 20 A^IB^{III} (A^I: Li to Cs; B^{III}: Al to Tl) compounds with the WIEN2k package [23], on the estimation of their formation energies $\Delta E^{o}_{AB,f,0GPa,0K}$ (kJ/mole) and on the estimation of the stabilizing pressure (GPa) for obtaining them in the DAC cell. In addition, we report also on high pressure in situ investigations up to 30 GPa with the DAC technique on the five thallides, such as LiTl, NaTl, KTl, RbTl, and CsTl.

2 Apparatus and Experimental Techniques

The apparatus and experimental techniques for studying the thallides LiTl, NaTl, KTl, RbTl, and CsTl in DAC technique are presented. The lithium, sodium, and potassium thallides were prepared at ambient pressure in a water-cooled copper boat by inductive heating of an equiatomic mixture of the constituent elements under argon. The rubidium and cesium A^ITl samples were synthesized by heating their equiatomic mixtures with thallium for 1 h at 500°C in corundum crucibles, which were placed in copper-sealed iron containers. In a DAC of Mao-Bell type, HP experiments were performed up to 30 GPa. Calibration of pressure was performed by adding 20% silver powder to the samples. With a CO_2 laser system, the DAC could be heated in site from both sides. Samples of RbTl and CsTl have been annealed for 2 h to about 100°C after HP treatment. By the Rietveld method, structural data were obtained. From literature, heat of formation data have been obtained for LiIn and for LiTl, at both 300 and 800 K.

The thallides LiTl, NaTl, and KTl were prepared at ambient pressure in a water-cooled copper boat by inductive heating of equiatomic mixtures with thallium under argon. Starting materials were lithium ingots (Alfa, nominal purity 99.9%), sodium and potassium metals (Koch-Light, 99.97%), and high-purity thallium (Johnson-Matthey, 99.997%). All manipulations were performed in a glove-box with high-purity argon. The reguli were cooled after each melting experiment, turned, and then molten several times again. Before last melting, the cold reguli were weighed back to control evaporation losses. After starting with alkali metal excess, the required amount of thallium was added, followed by several melting procedures. Due to the high vapor pressure of rubidium and cesium, samples of 1:1 composition with thallium were heated 1 h at 500°C in a corundum crucible, which was placed in copper-sealed iron container. At ambient pressure, these samples have been annealed for 2 h to about 100°C. From such samples, X-ray Guinier diffractograms (MoKα_1 radiation, quartz monochromator, 6–36° 2-theta, increment 0.04°, 20 s counting time per increment) were prepared for LiTl, NaTl, and KTl with isolated Tl_6-octahedra [22], Rb_8Tl_{11} [24] +3 Rb and CsTl with isolated Tl_6-octahedra [25].

In a piston-cylinder DAC of Mao-Bell type [26] (beveled 16-sided diamonds, culet: 300 μm, T301 hardened stainless steel gasket, thickness: 80 μm, hole: 130 μm), HP experiments were performed. No pressure-transmitting medium was applied. Calibration of pressure was performed by mixing 20% silver powder (5–8 μm; Johnson-Matthey Corp., #00785) to the samples. Specific volumes V/V_o of silver are tabulated as a function of pressure up to 90 GPa [27]. HP diffractograms were recorded either with AgKα radiation (λ = 0.5609 Å, Pd-filter) or with MoKα radiation (λ = 0.7107 Å, Zr-filter) in the theta range 3–13° (increment: 0.03°, typical counting time per increment: 600 s) with a collimator of 110 μm diameter. With a double-sided CO_2 laser system, both diamonds of the DAC could be heated. Two laser beams come to both sides of the diamond anvil and improve the temperature distribution. By the Rietveld method [28], structural data were obtained from the smoothed diffractograms, fixing the zero point of the diffractometer.

Formation enthalpies for A^IB^{III} compounds $\Delta H_{AB,f}$ which could be related to their stability are only known in a very few cases. Due to the experimental difficulties with these highly air-and moisture-sensitive A^IB^{III} compounds, such thermochemical data are mostly affected with very large uncertainties. A precise experimental procedure to obtain formation enthalpies $\Delta H_{AB,f,\ 800K}$ for LiIn with the NaTl- and for LiTl with the β-brass-type structure has been applied by Predel et al. [29]. In Predel's group, the Li-compounds as well as the pure components have been dissolved at 800 K in a calorimeter filled with liquid tin. For the compounds and their components, with samples of about 20 g 5–10 measurements, respectively, caloric experiments had been carried out [29]. The formation enthalpies $\Delta H_{AB,f,\ 800K}$ were obtained from the difference of the solution enthalpies of the components and the solution enthalpy of the compound [29]. The Miedema approach [30] was also applied to estimate enthalpies of formation $\Delta H_{AB,f,300}$ for LiIn and LiTl [29].

3 Computational Methods

The computational methods are presented. Total energy E^o calculations have been performed within the framework of DFT using the full-potential linearized augmented plane wave (LAPW) method embodied in the WIEN2k package. With the generalized gradient approximation (GGA), the exchange and correlation effects were treated using the functional of Wu and Cohen. Following a series of trials varying the $R_{mt}*K_{max}$ values between 7 and 11, a value of 9 was used. The maximal G in charge density Fourier expansion was 14. The total energies of the binary A^IB^{III} compounds, $E^o{}_{AB}$ and those of the constituents A^I and B^{III}, $E^o{}_A$ and $E^o{}_B$, were calculated for different volumes and fitting these data to Murnaghan's equation of state. After correction of the energies of formation $\Delta E^o{}_{AB,f}$ for overbonding, 6 of the 20 A^IB^{III} compounds have negative energies of formation and are therefore stable at zero pressure and temperature. From these fits, the equilibrium volumina and pressures were calculated, and, furthermore, the energies of formation at zero pressure and temperature via $\Delta E^o{}_{AB,f} = E^o{}_{AB} - E^o{}_A - E^o{}_B$ and the pressures needed for stabilizing the NaTl-type phases up to 30 GPa have been estimated.

Total energy calculations were performed within the framework of the DFT using the full-potential LAPW method embodied in the WIEN2k package [23]. With the GGA, the exchange and correlation effects were treated using the functional of Wu and Cohen [31]. During the whole calculations, the following muffin tin radii were used with constant values (a.u.): Li 2.1, Na 2.3, K 2.4, Rb 2.5, Cs 2.6; Al 2.3, Ga 2.3, In 2.4, and Tl 2.5. Total energy calculations E^o for the A^I metals (Li to Cs) were performed with the body-centered cubic structures (I–3 m). For Al, the face-centered cubic structure (Fm-3m), for Ga and In the tetragonal body-centered structure (I4/mmm), and for Tl the hexagonal-closed packed structure ($P6_3/mmc$) were used. The face-centered cubic structure (Fd-3m) was used for the A^IB^{III} compounds with the NaTl-type structure, the cubic primitive structure (Pm-3m) for the β-brass-type structure, and the body-centered tetragonal ($I4_1/amd$) and the tetragonal primitive (P4/mmm) structure for the two distorted structures. Lattice parameters were taken from "Pearson's Handbook of Crystallographic Data for Intermetallic Phases" [32, 33]. For energy calculations the tetragonal distorted NaTl-type structure ($I4_1/amd$), and for the CuAu-type structure (P4/mmm) lattice parameters were obtained by Rietveld analysis of diffractograms taken in our laboratory.

The total energies $E^o{}_{AB}$, $E^o{}_A$, and $E^o{}_B$ have been calculated for several volumes (e.g., typical changes (%): +20, +10, 0, −10, −20, −30, −40) and the volume–energy data were fitted to Murnaghan's equation of state [34]. From this fit, the equilibrium volume and pressure and the bulk moduli B_o (GPa) have been obtained. For a compound A^IB^{III}, the total enthalpy H^o is defined as $H^o = E^o + pV$ (p: pressure, V: volume). Thus, at zero pressure it is $H^o = E^o$. $E^o{}_{AB}$ is ground state–binding energy of the compound AB for one molecule. This value is derived from the total energy $E^o{}_{AB}$, $E^o{}_A$, and $E^o{}_B$: $E^o{}_{AB,f} = E^o{}_{AB} - E^o{}_A - E^o{}_B$. At zero temperature and pressure with $\Delta G^o{}_{AB,f} = \Delta H^o{}_{AB,f} + T \cdot \Delta S^o{}_{AB,f}$ the entropy term $T \cdot \Delta S^o{}_{AB,f}$ is zero; therefore, the free enthalpy of formation $\Delta G^o{}_{AB,f}$, the enthalpy of formation $\Delta H^o{}_{AB,f}$, and the energy of formation $\Delta E^o{}_{AB,f}$ are equal: $\Delta G^o{}_{AB,f} = \Delta H^o{}_{AB,f} = E^o{}_{AB,f}$. For the A^IB^{III} compounds, the heats of formation $\Delta H^o{}_{AB,f}$ at increasing pressures (0, 5, 10, 15, 20, 25, 30 GPa) were interpolated from the eosfit files of the WIEN2k package [23], which were fitted by the Murnaghan equation [34] for the energy-volume data of E_{AB}, E_A, and E_B, respectively. For Li to Cs, Al, Tl, NaTl-type, and β-brass-type structure, 4,000 k points (leading to 120 k points in the irreducible part of the Brillouin zone) were used. For Ga and In, the analogous values were 4,000/188 and 2,000/159. The maximal l value for partial waves inside the atomic spheres was 10, and the maximal G (magnitude of the largest vector) in charge density Fourier expansion was 14. The total energy E^o calculations have been performed with $R_{mt}*K_{max}$ values of 9 with $G_{max} = 14$. With increasing of $R_{mt}*K_{max}$ from 7 to 11, such calculations lead to more negative E^o values for the A^IB^{III} compounds.

However, the total energy differences between the NaTl- and the β-brass-type structure $\Delta E^o_{NaTl\text{-}\beta\text{-}brass}$ are not strongly affected by increasing $R_{mt}*K_{max}$, e.g., for KIn the following energy differences were obtained with $R_{mt}*K_{max} = 7$: –30.0, with 8: –29.3, with 9: –29.3, with 10: –29.6, and with 11: –29.6 (in mRyd). Although KIn is instable at zero pressure and temperature, the calculated energies of formation $\Delta E^o_{AB,f}$, e.g., for KIn are negative and increase slightly with increasing $R_{mt}*K_{max}$, e.g., for $R_{mt}*K_{max} = 7$, $\Delta E^o_{AB,f}$ is –10.4, for 8: –11.3, for 9: –11.8, for 10: –12.0, for 11: –11.9, for 12: –12.0 (in mRyd). Also the calculated energies of formation $\Delta E^o_{AB,f}$, for NaGa, KTl, RbIn, RbTl, CsTl become negative by WIEN2k calculation, implying that these compounds would be stable at zero pressure and temperature. This would lead to 12 stable A^IB^{III} compounds at zero pressure and energy. This is contradictory to the experimental facts that only six of them are stable at ambient conditions. Therefore, the energies of formation $\Delta E^o_{AB,f}$, calculated with $R_{mt}*K_{max} = 9$ and $G_{max} = 14$ have to be corrected for overbonding.

For LiIn and LiTl, the overbonding correction was fitted via the tabulated enthalpies of formation $\Delta H_{AB,f,800}$, measured by solution calorimetry and also calculated by the Miedema approach [29, 30]. For LiIn, the experimental energies of formation are $\Delta H_{LiIn,f,800} = -49$ kJ/mole (solution calorimetry) [29] and $\Delta H_{LiIn,f,300} = -46$ kJ/mol (Miedema approach) [29, 30]. The analogous energies of formation for LiTl are $\Delta H_{LiTl,f,800} = -40$ kJ/mole (solution calorimetry) [29] and $\Delta H_{LiIn,f,300} = -55$ kJ/mol (Miedema approach) [29, 30].

The applied overbonding corrections for the 20 A^IB^{III} compounds are shown in Fig. 1. They increase from LiAl with –1.0 to CsTl with –22.0 mRyd.

For LiTl (β-brass-type structure) and for LiIn and KTl (NaTl-type structure), the overbinding correction for the energies of formation $\Delta E^o_{AB,f}$ was also determined with the local spin density approximation (LSDA) by Perdew [35] and with the GGA by Perdew, Burke, and Ernzerhof ("Standard PBE") [36]. The overbinding corrections for LiIn, LiTl, and KTl obtained with GGA by Wu and Cohen [31] are 20% lower than those obtained with LSDA [35], but 15% higher than those obtained with "Standard PBE" GGA [36].

4 Crystal Structures

The crystal structures that were observed during the HP experiments with the DAC are presented here. The cubic NaTl- and the cubic β-brass-type structure are compared with each other. Two tetragonal distortions of both cubic phases are

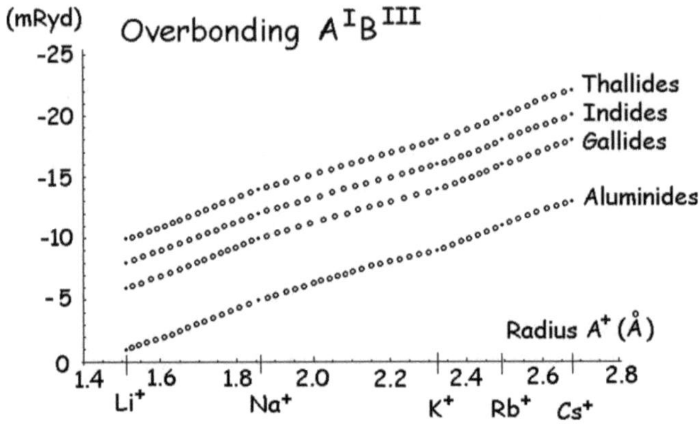

Fig. 1 Overbonding correction (mRyd) for A^I aluminides, gallides, indides, and thallides as function of the radius A^+ (Å) with Li^+ 1.509, Na^+ 1.858, K^+ 2.323, Rb^+ 2.493, Cs^+ 2.678

also discussed here. Both are obtained by a symmetry reduction in a group–subgroup relation "translationgleich" of index t3 from the cubic phases. The tetragonal distortion derived from the NaTl-type structure (Fd-3m) was found in 2002 by Ehrenberg et al. as low temperature phase of LiIn ($I4_1/amd$). The AuCu-type structure is the tetragonal distortion of the β-brass-type structure (P4/mmm) and was obtained in this investigation as HP phase of LiTl. In the 1930s of the last century, Zintl had observed this structure for LiBi and NaBi at ambient pressure.

4.1 NaTl-Type Structure

Figure 2 shows two views on the NaTl-type crystal structure, Pearson symbol cF16, space group Fd-3m with 8 f.u. per unit-cell. The face-centered cubic unit-cell of NaTl-type structure consists of eight small cubes, alternatively centered either with A^I or B^{III} atoms (Fig. 2a). Each of these centered sites has eight neighbors: four like and four unlike. The A^I and the B^{III} atoms lie on diamond sublattices, which are separated by a quarter of the body diagonal of the unit-cell. In the Zintl concept [1–6] of Na^+Tl^-, negatively charged Tl^- ion forms four equidistant and equiangled covalent bonds as they are observed in the diamond structure of tetrel elements C_{dia}, Si, Ge, α-Sn at ambient pressure (Fig. 2b). Up to now, six binary A^IB^{III} compounds crystallize in the NaTl-type structure. Zintl had characterized five compounds such as LiAl [1], LiGa [14], LiIn [14], NaIn [13], and NaTl [8] at ambient pressure, and one novel was observed by Evers and Oehlinger KTl at HP of 2 GPa [21].

4.2 β-Brass-Type Structure

Figure 3 shows a view on the β-brass-type crystal structure, Pearson symbol cP2, and space group Pm-3m with two f.u. per unit-cell. The cubes in the β-brass-type

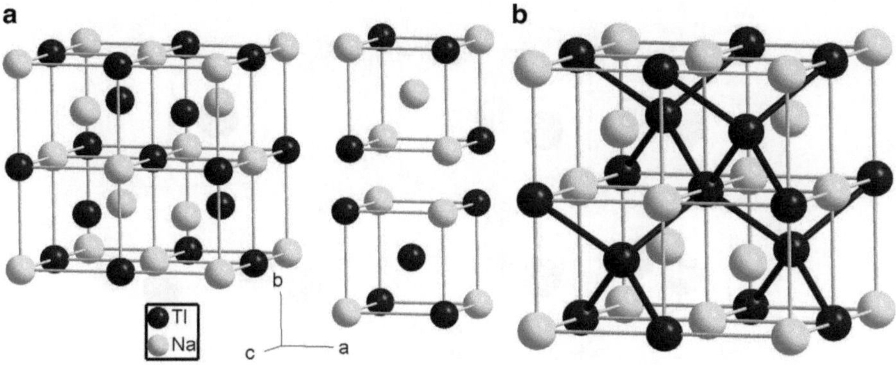

Fig. 2 View on the NaTl-type crystal structure, Pearson symbol cF16, space group Fd-3m. (**a**) The face-centered cubic arrangement of eight smaller cubes alternatively centered either with Tl or Na atoms. (**b**) In the Zintl concept of Na^+Tl^-, one electron is transferred from the electropositive Na atoms to the electronegative Tl atoms. By this electron transfer, the Tl^- ion forms four covalent bonds, as they are observed in the diamond structure of tetrel elements C, Si, Ge, and α-Sn

structure are either centered with A^I atoms or, after a shift of the origin with halve of the body diagonal, with B^{III} atoms. In the NaTl-type and in the β-brass alloy structure, each site has eight near neighbors in interpenetrating cubes, however, with different ordering. In contrast to the NaTl-type, in the β-brass-type structure there are eight equidistant A^IB^{III} contacts for every site. At ambient pressure, the A^IB^{III} compound LiTl crystallizes due to Zintl's investigation [14] with this structure. Schwarz et al. [20] prepared at HP LiIn with the β-brass-type structure. Since the lattice parameter of LiIn obtained at HP is approximately halved in comparison to the NaTl-type structure, eight cubic unit-cells of the β-brass-type structure have nearly the same volume as one of the normal pressure phase (Fig. 3).

4.3 Tetragonal Distorted NaTl-Type Structure

Due to low temperature investigations of Ehrenberg et al. [19], LiIn transforms at 170 K at ambient pressure into a tetragonal distorted NaTl-type structure, Pearson symbol tI8, space group $I4_1/amd$ with four f.u. per unit-cell. The symmetry is reduced in a reversible group–subgroup "translationengleich" transition of index 3 (t3). The tetragonal distortion is with 0.5% [19] very small. The first Li–In coordination shells are only slightly distorted. Distances are not affected, but the ideal tetrahedral angle of 109.47° splits into those with 109.20° and 109.61° [19]. Neglecting this small distortion, the low temperature structure of LiIn (Fig. 3) in this distorted NaTl-type structure contains alternatively centered interpenetrating cubes that have four like and four unlike neighbors. Figure 4 shows a view on the low temperature structure of LiIn.

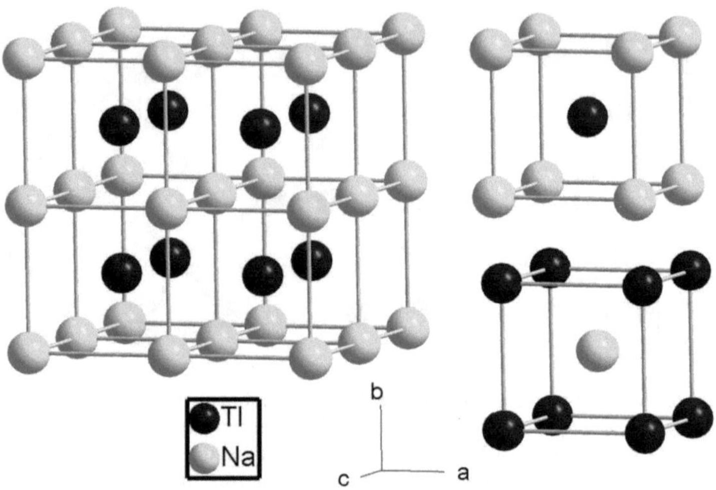

Fig. 3 View on eight unit-cells of the β-brass alloy crystal structure, Pearson symbol cP2, space group Pm-3m. The unit-cells are either centered with A atoms or after a shift of halve of the body diagonal with B^{III} atoms. Every site has eight A^{I}–B^{III} contacts equidistant and equiangled

4.4 AuCu-Type Structure

The AuCu-type structure, Pearson symbol tP4, space group P4/mmm with two f.u. per unit-cell ($z = 2$), can be considered as a tetragonal distortion of the β-brass alloy structure (CsCl-type structure) ($z = 1$). The transformation of the β-brass-type structure (Pm-3m) into the tetragonal distorted structure with $z = 1$ (P4/mmm) is also a symmetry reduction "translationgleich" of index 3 (t3). Zintl and coworkers determined the tetragonal distorted β-brass alloy structure with $z = 1$ for LiBi [37] and for NaBi [38]. In Sect. 5.4, the structural data for LiBi and NaBi will be transformed into those for the AuCu-type structure ($z = 2$) to compare it with the high pressure phase determined here for LiTl. In Fig. 5, the AuCu-type structure and the tetragonal distorted β-brass-type structure are shown.

The normal pressure phases of KTl [22] and CsTl [25] with isolated Tl_6^- octahedral, which are easily transformed at HP, are not considered here.

5 Results and Discussion

The results of the WIEN2k calculations are presented for aluminides, gallides, indides, and thallides. In addition, experimental results of in situ HP investigations in the DAC on thallides are presented here. The WIEN2k calculations show that for three $A^{I}B^{III}$ compounds with small radius ratio (r_A:r_B from 1.0 to 1.1) LiAl, LiGa, and NaTl, the NaTl-type structure could be transformed into the β-brass-type structure in the pressure range up to 40 GPa. These compounds simulate the

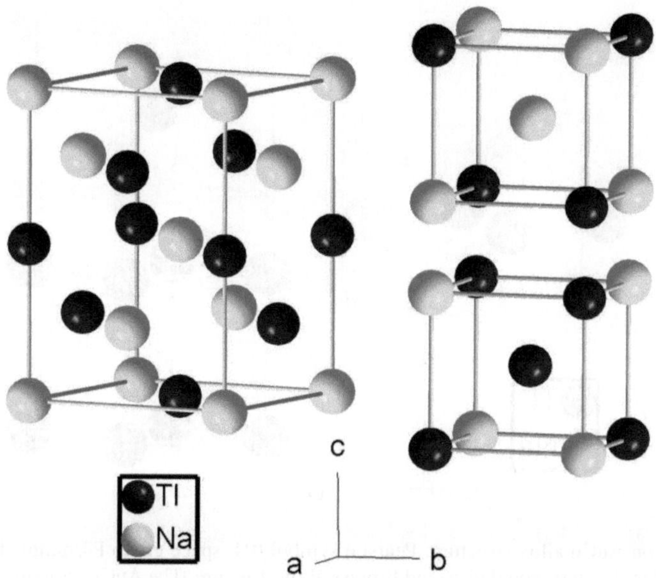

Fig. 4 View on the low temperature phase of LiIn with the tetragonal distorted NaTl-type crystal structure, Pearson symbol tI8, space group $I4_1/amd$. The tetragonal distortion is with 0.5% small. The body-centered cubic arrangement of this structure contains interpenetrating cubes alternatively centered either with A^I or with B^{III} atoms with four like and four unlike neighbors. The transition of α-Sn (Fd-3m) into β-Sn ($I4_1/amd$) at 15° follows the same route of symmetry reduction, as has been observed for the In atoms in LiIn. The transition in Sn is also "translationengleich" of index 3, but the tetragonal distortion in β-Sn is much stronger than in LiIn. Due to the strong compression of the *c*-lattice parameter in β-Sn, the coordination of the Sn atoms is increased from four to six with a density increase of about 20%, which is about 40 times stronger than in LiIn

behavior of LiTl at ambient pressure. For higher $r_A{:}r_B$ ratios, the NaTl-type structure is more stable than the β-brass-type structure. In a DAC diffractograms of LiTl, NaTl, KTl, RbTl, and CsTl were obtained and then analyzed with the Rietveld technique. For 19 A^IB^{III} compounds with the NaTl-type structure, the lattice parameters have been determined and the near-neighbor diagram for these phases has been derived. The near-neighbor diagram, introduced by Pearson in 1968, is used to get information on the bonding characteristics for intermetallic compounds. For the NaTl-type structure, it is now extended from radius ratio $r_A{:}r_B$ of 1.15 for five at that time actual phases known by Pearson to now 19 phases with such a ratio up to 1.90.

5.1 Aluminides LiAl, NaAl, KAl, RbAl, and CsAl

The binary phase diagrams of the 1:1 alkali aluminides show at 1 bar two different types of phase diagrams [39]. The congruent melting of LiAl at 700°C allows an easy preparation from a melt with 1:1 composition. The phase diagrams

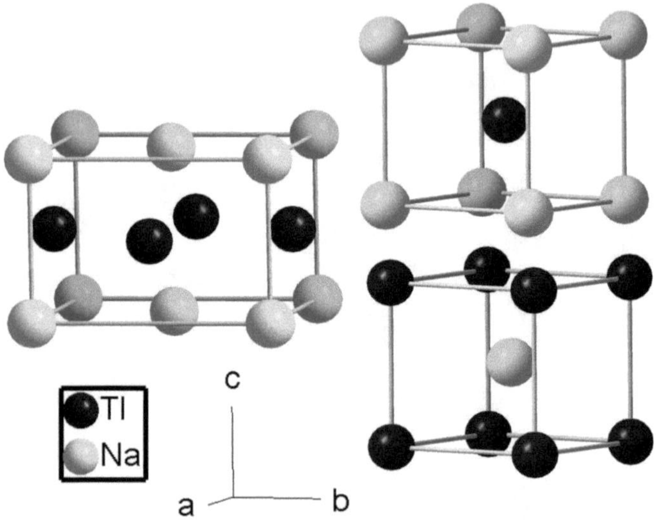

Fig. 5 View on AuCu alloy structure, Pearson symbol tP4, space group P4/mmm. This structure can be considered as tetragonal distorted β-brass alloy structure. The AuCu structure contains two f.u. The tetragonal distortion is stronger than in the low temperature phase LiIn. In the AuCu-type structure, there are tetragonal distorted interpenetrating cubes, and are either centered with A^I or B^{III} atoms. Each of these interpenetrating cubes contains only one formula unit AuCu and can be considered as an own structure, Pearson symbol tP2, space group P4/mmm

Na-Al, K-Al, Rb-Al, and Cs-Al show that these elements are practically immiscible in the liquid state near the melting point of Al at 660°C [39]. Therefore, the compounds NaAl, KAl, RbAl, and CsAl could not be prepared up to now from a melt at 1 bar.

In Table 1, the calculated total energy E^o(Ryd), the volume V_o (Å^3), both for one f.u., and the bulk moduli B_o (GPa) for the aluminides LiAl, NaAl, KAl, RbAl, and CsAl with the NaTl- and the β-brass-type structure at zero pressure and temperature are summarized.

For the stable compound LiAl, the lattice parameter for the NaTl-type structure ($z = 8$) $a_{0K,0GPa}$ is calculated via the cubic root of $8{\cdot}V_o$. With $V_o = 31.71$ Å^3 (Table 1), $a_{0K,0GPa} = 6.33$ Å is calculated. The tabulated lattice parameters [32, 33] for LiAl are $a_{300K,1bar}$: 6.3667, 6.3703, 6.3757, in Å, with an average value of $a_{300Kaverage} = 6.371$ Å, leading to a unit-cell volume of $V_{average300K} = 258.6$ Å^3. In order to compare both lattice parameters, one has to correct the experimental lattice parameter of LiAl for thermal expansion.

During this investigation, the lattice parameters of LiIn with the NaTl-type structure have been determined from Guinier diffractograms with CuKα_1 radiation at 300 and 200 K and for the tetragonal distorted NaTl-type structure at 100, 50, 30, and 10 K (Table 7). For LiIn with a cubic unit-cell volume at 300 K of 313.7 Å^3, there is on cooling to 10 K a volume decrease of $\Delta V = -5.63$ Å^3 for the tetragonal phase. For LiAl with a volume of 258.6 Å^3 at 300 K, assuming

Table 1 Calculated total energy E^o (Ryd), volume V_o(Å^3) for one f.u., and bulk moduli B_o (GPa) of aluminides LiAl, NaAl, KAl, RbAl, and CsAl with the NaTl- and the β-brass-type structure at zero pressure and temperature

	NaTl-type structure			β-Brass-type structure		
	E^o(Ryd)	V_o(Å^3)	B_o(GPa)	E^o(Ryd)	V_o(Å^3)	B_o(GPa)
LiAl	–500.23568	31.71	46	–500.22393	30.66	42
NaAl	–809.73778	41.90	32	–809.70616	46.19	25
KAl	–1,688.82481	54.44	22	–1,688.78208	66.12	14
RbAl	–6,447.04091	60.49	20	–6,447.00136	74.48	12
CsAl	–16,064.10790	65.27	22	–16,064.06805	75.66	15

proportionality of thermal expansion between LiIn and LiAl, a decrease of $\Delta V = -4.6$ Å^3 is calculated (volume of 254.0 Å^3 at 10 K, 1 bar). The experimental lattice parameter of LiAl ($a_{300\text{K},1\text{bar}} = 6.371$ Å) is then decreased to $a_{10\text{K},1\text{bar}} = 6.33$ Å, which is now in total agreement with the calculated lattice parameter (WIEN2k) of $a_{0\text{K},0\text{bar}} = 6.33$ Å. With the local density approach (LDA), Christensen calculated in 1985 a $_{0\text{K},0\text{bar}} = 6.328$ Å for LiAl [18].

Inspection of Table 1 shows that LiAl with the NaTl-type and the β-brass-type structure have the highest bulk moduli B_o (46, 42, respectively) of all the 1:1 alkali aluminides. LiAl is therefore the hardest material of these aluminides. Table 1 indicates also that the bulk moduli of the phase with the β-brass-type structure have lower B_o values than those for NaTl-type structure and are, therefore, more ductile. Thus, the phases with the β-brass type seem to have more alloy character than those with the NaTl-type structure with partly covalent bonding in the Tl^- net. Christensen [18] calculated in 1985 the bulk modulus of LiAl with the NaTl-type structure to 45.0 GPa at the theoretical equilibrium volume. Interestingly, the bulk moduli B_o of CsAl (22, 15 GPa) are higher than those of RbAl (20, 12 GPa) (Table 1).

The differences in energy ΔE^o and volume ΔV_o between the phases with NaTl- and β-brass-type structure (Table 1) are strongly dependent on the size of the alkali metals. In Fig. 6, the energy difference ΔE (Ryd) between the NaTl- and the β-brass type as function of the radii (Å) from Li to Cs is shown; in Fig. 7, the volume difference ΔV (a.u.^{3}) between both structures is plotted.

Interestingly, the β-brass-type phase of LiAl is the only phase of the 1:1 alkali aluminides which is lower in volume V_o (30.66 Å^3, Table 1, 206.93 a.u.^{3}, with 1 a.u. = 0.529177 Å) than the NaTl-type structure (31.71 Å^3, Table 1, 214.01 a.u.^{3}), although the energy E^o of the first (−5000.22393 Ryd, Table 1) is –11.76 mRyd lower than that of the second (−500.23568 Ryd, Table 1). Christensen calculated this energy difference −12.6 mRyd [18]. However, as can be seen in Fig. 7, the energy difference between the NaTl- and the β-brass type in the series of aluminides is the lowest for LiAl. Therefore, applying high pressure on LiAl one could transform the NaTl-type structure. In Fig. 8, the total energies E^o (Ryd) and the volumes V (a.u.^{3}) for LiAl with the NaTl- and the β-brass type are plotted as function of the unit-cell volume.

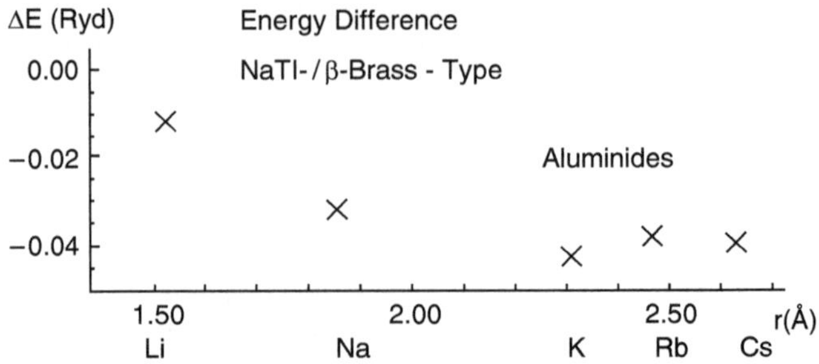

Fig. 6 Energy difference ΔE^o (Ryd) between the NaTl- and the β-brass-type structures of LiAl, NaAl, KAl, RbAl, and CsAl as function of the radii (Å) for the alkali metals (CN 8)

At about 155 a.u.^{3}, the total energies E^o for both phases are equal, thus allowing coexistence of both phases. From the fit of the volume–energy data with Murnaghan's equation of state [34], a pressure of 25 GPa can be estimated for the transformation of the NaTl- into the β-brass-type structure of LiAl.

For NaAl, the difference in energy is too high that a HP transformation in the range up to 30 GPa into the β-brass-type structure seems to be successful. As it has been already mentioned in Sect. 3, the energy of formation $\Delta E^o_{AB,f}$ at zero temperature and pressure can be calculated from the total energies $E^o{}_{AB}$, $E^o{}_A$, and $E_B{}^o$ with the $\Delta E^o_{AB,f} = E^o{}_{AB} - E^o{}_A - E^o{}_B$. In Table 2, the energies of formation for NaAl $\Delta E^o{}_{f,NaAl}$ (Ryd) with increasing pressure are summarized, without applying an overbonding correction. From inspection of Table 2, it is evident that the reaction Na(s) + Al(s) → NaAl(s) at zero temperature and pressure leads to a positive energy of formation $\Delta E^o{}_{NaAl,f} = +19.04$ mRyd, which converts into $\Delta E^o_{NaAl,f} = +25$ kJ/mole. After correction for overbonding (+5.0 mRyd for NaAl), $\Delta E^o{}_{NaAl,,f,\ corr} = +19.04 + 5.0 = +24.04$ mRyd, which converts into $\Delta E^o{}_{NaAl,,f,corr} = +32$ kJ/mole. The reaction to form NaAl from the constituent metals is therefore endergonic at zero pressure and temperature. However, with increasing pressure the positive $E^o{}_{NaAl,f}$ values decrease (Table 2) until at 25 GPa the reaction is exergonic with $\Delta E^o{}_{NaAl,f} = -5.13 + 5.0 = -0.13$ mRyd (Table 2), after overbonding correction. The reason for the decrease of the positive values of $\Delta E^o{}_{NaAl,f}$ in the pressure range up to 25 GPa is the stronger decrease of $E^o{}_{Na}$ in comparison to $E^o{}_{Al}$ and to E^o_{NaAl}.

At 1 bar, NaAl cannot be prepared from the liquid elements, because sodium and aluminum are practically immiscible at the melting point of Al [39]. Therefore, with a synthesis in DAC with double-sided laser heating at about 25 GPa, there could be a chance to synthesize NaAl with the NaTl-type structure.

The estimated energies of formation $\Delta E^o{}_{ABf,0K,0GPa,}$ of NaTl-type phases LiAl to CsAl and the estimated pressure (GPa) for stabilizing the NaTl-type structure in NaAl to CsAl are summarized in Table 3.

For 1:1 alkali aluminides, LiAl is the only stable phase with the NaTl-type structure at 1 bar. The estimated energy of formation (WIEN2k) at zero pressure and temperature is $\Delta E^o{}_{LiAl,f,0K,0GPa}$ −33 kJ/mole. With increasing metal radius from Na to Cs, the

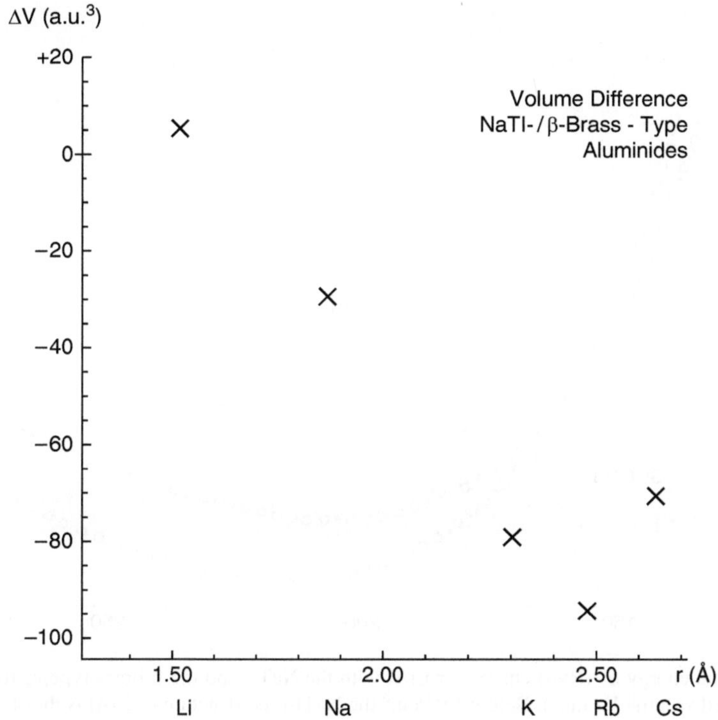

Fig. 7 Volume difference ΔV_o (a.u.^{3}) between the NaTl- and the β-brass-type structures of LiAl, NaAl, KAl, RbAl, and CsAl as function of the radii (Å) for the alkali metals (CN 8)

endergonic formation energies increase drastically (NaAl +32, KAl +70, RbAl +90, CsAl +100 kJ/mole; Table 3). The estimated stabilization pressure for the alkali aluminides with the NaTl-type structure from NaAl to CsAl lies in the range between 25 and 30 GPa (Table 3).

5.2 *Gallides LiGa, NaGa, KGa, RbGa, and CsGa*

The binary phase diagrams of the 1:1 alkali gallides show at 1 bar nearly the analogous behavior as the 1:1 alkali aluminides. The congruent melting of LiGa at 730°C [39] allows only the preparation of this compound from a melt. The phase diagrams Na-Ga, K-Ga, Rb-Ga and Cs-Ga indicate that these elements are practically immiscible in the liquid state [39]. Therefore, the compounds NaGa, KGa, RbGa, and CsGa cannot be prepared up to now from a melt at 1 bar.

In Table 4, the calculated total energy E^o(Ryd), the volume V_o (Å^3), both for one f.u., and the bulk moduli B_o (GPa) for the gallides LiGa, NaGa, KGa, RbGa, and

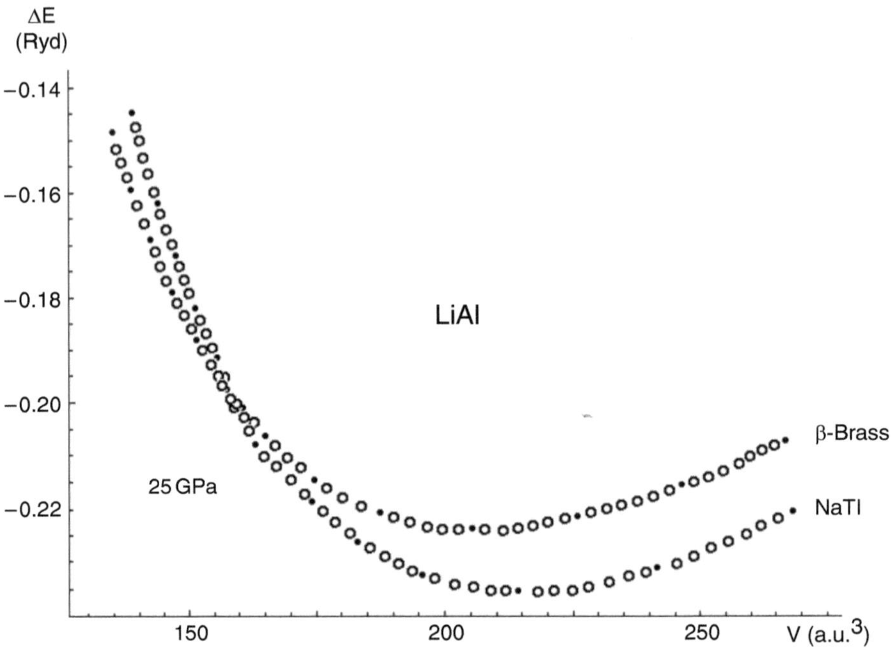

Fig. 8 Total energy E^{o} (Ryd) curves for LiAl with the NaTl- and the β-brass-type as function of the unit-cell volume V (a.u.^{3}). Below 155 a.u.^{3} the NaTl-type structure of LiAl is the stable phase

Table 2 Estimation of the energy of formation $\Delta E^{o}_{f,NaAl}$ (Ryd) for NaAl with the NaTl-type structure as function of the pressure p (GPa) up to 40 GPa, not corrected for overbonding

p (GPa)	Na(s) (Ryd)	Al(s) (Ryd)	NaAl(s) (Ryd)	$\Delta E^{o}_{f,NaAl,0K}$ (Ryd)
40	−324.42771	−485.20887	−809.65118	−0.01460
35	−324.43806	−485.21369	−809.66350	−0.01160
30	−324.44903	−485.21835	−809.67588	−0.00849
25	−324.46026	−485.22276	−809.68815	−0.00513
20	−324.47199	−485.22689	−809.70037	−0.00149
15	−324.48423	−485.23061	−809.71226	+0.00257
10	−324.49740	−485.23373	−809.72342	+0.00771
5	−324.51011	−485.23599	−809.73290	+0.01319
0	−324.51991	−485.23690	−809.73778	+0.01904

CsGa with the NaTl- and the β-brass-type structure at zero pressure and temperature are summarized.

The differences in energy ΔE^{o} and volume ΔV_{o} between the phases with NaTl- and β-brass-type structure (Table 4) are strongly size dependent. For LiGa, the difference in energy is $\Delta E^{o} = -10.2$ mRyd and the difference in volume $\Delta V_{o} = -0.26$ Å^{3}, but for CsGa such values are −20.7 mRyd and −12.15 Å^{3}, respectively. Christensen calculated for LiGa the energy difference between the NaTl- and the β-brass-type structure to −8.6 mRyd [18].

Table 3 Estimated energy of formation $\Delta E^{o}_{f,AB,0GPa}$ (kJ/mole) for LiAl to CsAl, and estimated pressure (GPa) for NaTl phases stabilized in NaAl to CsAl

A^IB^{III} phase with NaTl-type structure	LiAl	NaAl	KAl	RbAl	CsAl
$\Delta E^{o}_{f,AB,0GPa}$ (kJ/mole)	−33	+32	+70	+90	+100
Stabilization pressure (GPa)	0	25	30	30	30

Inspection of Table 4 shows that LiGa with the NaTl-type and the β-brass-type structure has the highest bulk moduli B_o (45, 41 GPa, respectively). LiGa is therefore the hardest material of these alkali gallides. Christensen calculated the bulk modulus B_o of LiGa to 43.3 GPa [18]. Table 4 shows also that the bulk moduli B_o of the phases with the β-brass type have lower values than those with NaTl-type structure. Therefore, the first are more ductile than the second. Interestingly, the bulk modulus B_o of CsGa with the NaTl-type structure (22 GPa) is higher than that of RbGa (19 GPa), although one would expect the opposite.

For the ambient pressure stable compound LiGa, the lattice parameter $a_{0K,0GPa}$ for the NaTl-type structure ($z = 8$) is calculated from the cubic root of $8{\cdot}V_o$. With $V_o = 29.45$ Å^3 (Table 1), one calculates $a_{0K,0GPa} = 6.18$ Å. The tabulated lattice parameters [32, 33] for LiGa are $a_{300K,1bar}$: 6.195, 6.22, 6.150, in Å, with an average value of $a_{300Kaverage} = 6.188$ Å, leading to a unit-cell volume of $V_{average300K} =$ 236.95 Å^3. After correction for thermal expansion, the experimental lattice parameter of LiGa ($a_{300K,1bar} = 6.188$ Å) is then decreased to $a_{10K,1bar} = 6.15$ Å, which is now in very good agreement with the calculated lattice parameter (WIEN2k) of $a_{0K,0bar} = 6.18$ Å. Christensen had calculated the lattice parameter of LiGa to $a_{0K,0GPa} = 6.405$ Å [18]. The large deviation of this calculated value (6.405 Å) to the experimental one (6.195 Å), Christensen commented that he had "Ga 3d treated as frozen states (causes overestimate of a_{theor})" [18].

In Fig. 9, the total energies E^o (Ryd) and the volumes V (a.u.^{3}) for LiGa with the NaTl- and the β-brass type are plotted as function of the unit-cell volume. At about 140 a.u.^{3}, the total energy E^o curves touch each other. The phases of LiGa with the NaTl- and with the β-brass-type structure coexist in this case. With an energy difference of −10.2 mRyd (= −13 kJ/mole, Table 4), the transformation pressure is approximately 25 GPa. The difference in energy is for NaGa (−21.7 mRyd = −28 kJ/mole; Table 4) too high that NaGa can be transformed up to 30 GPa into the β-brass-type structure.

The estimated energies of formation $\Delta E^{o}_{f,AB,0K,0GPa}$ for LiGa to CsGa and the estimated pressure (GPa) for NaGa to CsGa stabilizing the NaTl-type structure are summarized in Table 5. For alkali gallides, LiGa is the only stable phase with the NaTl-type structure at 1 bar. The estimated energy of formation (WIEN2k) for LiGa at zero pressure and temperature is $\Delta E^{o}_{f,LiAl,f,0K,0GPa} = -61$ kJ/mole. With increasing metal radius from Na to Cs, the endergonic formation energies increase drastically. The estimated stabilization pressure for the alkali gallides NaGa to CsGa with the NaTl-type structure lies in the range between 5 and 15 GPa (Table 5).

Table 4 Calculated total energy E^o(Ryd), volume V_o(Å^3) for one f.u., and bulk modulus B_o(GPa) of gallides LiGa, NaGa, KGa, RbGa, and CsGa with the NaTl- and the β-brass-type structure at zero pressure and temperature

	NaTl-type structure			β-Brass-type structure		
	E^o(Ryd)	V_o(Å^3)	B_o(GPa)	E^o(Ryd)	V_o(Å^3)	B_o(GPa)
LiGa	−3,902.35348	29.45	45	−3,902.34327	29.71	41
NaGa	−4,211.85705	40.13	30	−4,211.83539	44.45	26
KGa	−5,090.94716	52.60	20	−5,090.91811	63.21	16
RbGa	−9,849.16447	59.05	19	−9,849.13731	70.36	14
CsGa	−19,466.23272	63.99	22	−19,466.21205	76.14	14

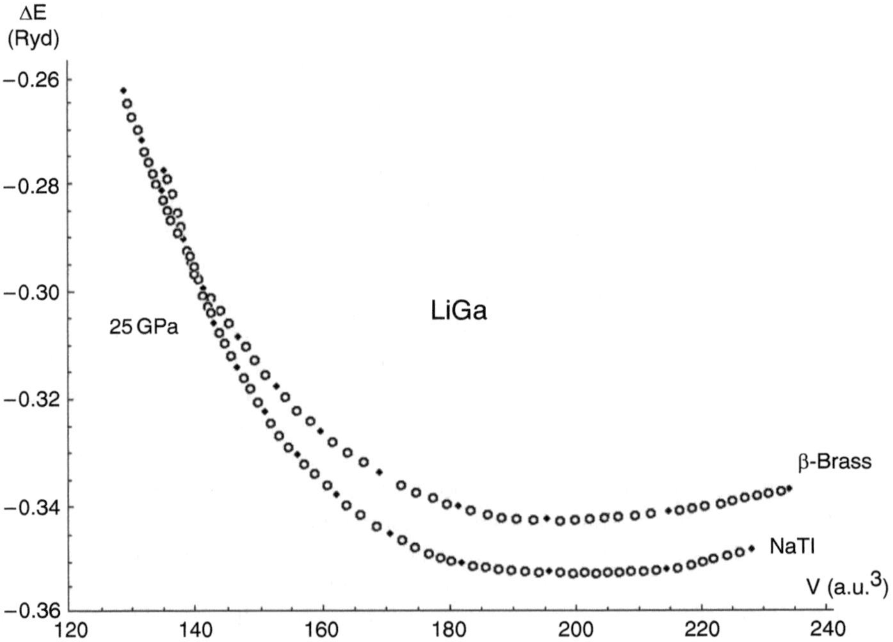

Fig. 9 Total energy E^o (Ryd) curves for LiGa with the NaTl- and the β-brass-type as function of the unit-cell volume V (a.u.^{3}). At 140 a.u.^{3}, the total energy with about 25 GPa pressure, the total energy E^o curves touch each other. At this pressure, the NaTl- and the β-brass-type structure of LiGa are in equilibrium

5.3 Indides LiIn, NaIn, KIn, RbIn, and CsIn

The binary phase diagrams of the 1:1 alkali indides show at 1 bar different types of phase diagrams [39]. As observed for LiAl and LiGa, also LiIn melts congruently [632(5)°C]. For NaIn, less stable than LiIn, no dystectical, but peritectical melting at 345(4)°C is found. No compound KIn is stable in the K-In-system, but there is a miscibility gap near 1:1 composition. Such a behavior occurs also for the Rb-In and the Cs-In system.

Table 5 Estimated energy of formation $\Delta E^{o}_{f,AB,0K,0GPa}$ (kJ/mole) for LiGa to CsGaand estimated pressure (GPa) for NaTl-type phases stabilized in NaGa to CsGa

$A^{I}B^{III}$ phase with NaTl-type structure	LiGa	NaGa	KGa	RbGa	CsGa
$\Delta E^{o}_{f,AB,0K,0GPa}$ (kJ/mole)	−61	+2	+36	+64	+63
Stabilization pressure (GPa)	0	5	10	15	10

In Table 6, the calculated total energy E^{o}(Ryd), the volume V_{o} (Å^3), both for one f.u., and the bulk moduli B_{o} (GPa) for the indides LiIn, NaIn, KIn, RbIn, and CsIn with the NaTl- and the β-brass-type structure at zero pressure and temperature are summarized.

As it has been also observed for aluminides and gallides, for indides the differences in energy ΔE^{o} and volume ΔV_{o} between the phases with NaTl- and β-brass-type structure (Table 4) are strongly size dependent, having in mind that LiIn with β-brass-type structure – due to this investigation – is more stable than that with the NaTl-type phase (−1.1 mRyd). Christensen, however, calculated in 1985 for LiIn the opposite that NaTl-type structure −1.3 mRyd more stable than the β-brass-type structure [18]. Due to our results, LiIn with NaTl-type structure is then high temperature phase at 1 bar, the β-brass modification low temperature phase. With a volume $V_{o} = 37.69$ Å^3, the β-brass phase of LiIn has also a lower volume than the NaTl-type phase (39.46 Å^3 with $\Delta V_{o} = -1.77$ Å^3 = −4.3%). However, for NaIn to CsIn the differences in energy and in volume between the NaTl- and the β-brass-type structure increase. For NaIn, $\Delta E^{o} = -16.8$ mRyd and in volume $\Delta V_{o} = -2.54$ Å^3, but for CsIn these values are −26.1 mRyd and −14.93 Å^3, respectively (Table 6). Christensen has calculated the energy difference between the NaTl- and β-brass-type structure for NaIn to −14.2 mRyd [18].

Inspection of Table 6 shows that LiIn phases with the NaTl-type and the β-brass-type structure have the highest bulk moduli B_{o} (36, 35 GPa, respectively) and are therefore the hardest materials of these 1:1 alkali indides. NaIn has bulk moduli B_{o} (28, 25 GPa, respectively, for the NaTl- and the β-brass-type structure). Christensen calculated a bulk modulus B_{o} of 34.7 GPa for LiIn and for 31.1 GPa for NaIn with the NaTl-type structure [18]. Schwarz et al. derived experimentally the bulk moduli B_{o} of LiIn [39(5), 46(5) in GPa, respectively, for the NaTl- and the β-brass-type structure] [20].

For LiIn with NaTl-type structure, the lattice parameter $a_{0K,0GPa}$ is calculated from the cubic root of $8 \cdot V_{o}$ (Table 6). With $V_{o} = 39.46$ Å^3, a lattice parameter of $a_{0K,0GPa} = 6.81$ Å for LiIn is obtained. The tabulated lattice parameters [32, 33] for LiIn are $a_{300K,1bar}$: 6.792, 6.786, 6.7926, in Å, with an average value of $a_{300Kaverage} = 6.790$ Å, leading to a unit-cell volume of $V_{average300K} = 313.05$ Å^3 (with 39.13 Å^3, $z = 1$, 300 K). After correction for thermal expansion (38.52 Å^3, 10 K, $z = 1$, slightly tetragonal distorted NaTl-type structure for LiIn, Table 7) (38.52–39.13 = −0.61 Å^3) with 8·(−0.61) = −4.88 Å^3 (313.05–4.88 = 308.17 Å^3), a lattice parameter of $a_{10K,1bar} = 6.75$ Å is obtained for LiIn, in comparison with the WIEN2k value of $a_{0K,0GPa} = 6.81$ Å. Christensen had calculated a lattice parameter of $a_{0K,0GPa} = 7.093$ Å for LiIn [18]. The large deviation to the experimental

Table 6 Calculated total energy E^{o} (Ryd), volume V_o(Å^3) for one f.u., and bulk moduli B_o (GPa) of indides LiIn, NaIn, KIn, RbIn, and CsIn with the NaTl and the β-brass-type structure at zero pressure and temperature

	NaTl-type structure			β-Brass-type structure		
	E^{o}(Ryd)	V_o(Å^3)	B_o(GPa)	E^{o}(Ryd)	V_o(Å^3)	B_o(GPa)
LiIn	−11,780.19109	39.46	36	−11,780.19223	37.69	35
NaIn	−12,089.71673	49.07	28	−12,089.69992	51.61	25
KIn	−12,968.81673	61.69	18	−12,968.78701	70.80	16
RbIn	−17,727.03623	68.20	16	−17,727.00640	79.47	14
CsIn	−27,344.10551	71.70	20	−27,344.07946	86.63	13

value (a_{exp} = 6.786 Å) Christensen commented that he had "In treated as 'frozen core' states (causes overestimate of a_{theor})" [18].

The tabulated lattice parameters [32, 33] for NaIn are $a_{300K,1bar}$: 7.332 and 7.297, in Å. The value of 7.297 Å has been determined by Zintl in 1933 [13], using a wavelength at that time of lower precision and may here be omitted. In the "Handbook for Intermetallic Phases," first printing 1985 [32], there is a lattice parameter of NaIn (7.332 Å) which is given twice from the measurements performed in 1970 and 1979 [32, 33]. With a value of $a_{300Kaverage}$ = 7.332 Å, a unit-cell volume of $V_{average300K}$ = 394.16 Å^3 is obtained. After correction for thermal expansion, assuming proportionality between LiIn (−4.88 Å^3 at 313.05 Å^3, see above), a thermal correction of $-\Delta V = -6.1$ Å^3 is calculated for NaIn with 394.16 Å^3 at 300 K. This leads to a corrected experimental lattice parameter at 10 K of NaIn of $a_{10K,1bar}$ = 7.29 Å. This value is in good agreement with the WIEN2k calculated value of $a_{0K,0GPa}$ = 7.30 Å. Christensen's lattice parameter for NaIn of $a_{0K,0GPa}$ = 7.518 Å shows again a large deviation to his experimental value of 7.321 Å [18].

The differences in energy ΔE^{o} and volume ΔV_o between the phases with NaTl- and β-brass-type structure (Table 6) are strongly dependent on the size of the alkali metals, as also has been observed for the aluminides and gallides. Inspection of Table 6 and Fig. 10 shows that a structural change takes place for LiIn.

In Fig. 9, the total energies E^{o} are plotted as function of the volume. At 7.7 GPa, the NaTl phase has a volume of 230 a.u.^{3} (=34.08 Å^3), and the β-brass phase one of 217 a.u.^{3} (=32.16 Å^3). These values (34.08 and 32.16 Å^3) can also be used to calculate the lattice parameters at 7.5 GPa and zero temperature. The cubic root of 32.16 Å^3 ($z = 1$) is the lattice parameter a of the β-brass phase: a_{0K} = 3.18 Å. For the NaTl phase, one has to calculate the lattice parameter from 8·34.08 Å^3 ($z = 8$). The cubic root is a_{0K} = 6.48 Å. These values can be compared with the data of Schwarz et al. [20] who performed high pressure experiments (1998) on LiIn up to 18 GPa obtaining the β-brass phase from the NaTl phase. From their pressure–volume curve fitted to the experimental data, one can derive the volume, e.g., at 7.5 GPa. One obtains for the NaTl phase 34.4 Å^3 and for the β-brass phase 33.3 Å^3. From these data, the lattice parameters at 7.5 GPa and 300 K a_{300K} = 3.22 Å for the β-brass phase and a_{300K} = 6.50 Å for the NaTl phase are derived. These data have to be corrected for thermal expansion with $\Delta V = -0.73$ Å^3 for $z = 1$ and

Table 7 Space group, lattice parameter, volume for one f.u., and deviation from the ideal *c/a* ratio for LiIn obtained with Guinier diffractograms with CuKα1 radiation in the temperature range between 300 and 10 K. The volume difference between 10 and 300 K is $\Delta V = -0.73$ Å^3

T(K)	Space group	*a* (Å)	*c* (Å)	$c/(a\sqrt{2})$	*V* (Å)^3
10	$I4_1/amd$	4.7704(2)	6.7709(4)	1.0036	38.52(1)
30	$I4_1/amd$	4.7708(2)	6.7713(4)	1.0036	38.53(1)
50	$I4_1/amd$	4.7720(2)	6.7721(4)	1.0035	38.55(1)
100	$I4_1/amd$	4.7759(2)	6.7744(5)	1.0030	38.63(1)
200	Fd-3m	6.7809(3)			38.97(1)
300	Fd-3m	6.7970(3)			39.25(1)

a $\Delta V = -0.73 \cdot 8 = -5.84$ Å^3 for $z = 8$ (Table 7). The thermal corrected values of the lattice parameters of Schwarz et al. [20] are $a_{10K,7.5GPa} = 3.19$ Å and $a_{10K,7.5GPa} = 6.46$ Å which are in very good agreement with the values at 0 K $a_{0K,7.5GPa} = 3.18$ Å and $a_{0K,7.5GPa} = 6.48$ Å, calculated by WIEN2k.

However, for LiIn two other structures lie near to the energy minimum of –11,780.19223 Ryd (Table 6). These are the tetragonal distortions of the NaTl-type and of the β-brass-type structure. The tetragonal distortion the cubic NaTl-type structure was observed in 2002 by Ehrenberg et al. [19] as low temperature phase (see Sect. 4.3). The tetragonal distortion of the β-brass-type structure was obtained during our DAC experiments with LiTl (see Sect. 5.4). The transition temperature for cubic LiIn with the NaTl-type structure is 170(10) K [19]. In order to determine the thermal expansion, which is needed for comparison of the lattice parameters of A^IB^{III} phases at room temperature with those calculated by WIEN2k at zero temperature and pressure, cooling experiments down to 10 K have been performed on LiIn (Rotter, 2010, Department of Chemistry, LMU Munich, unpublished results) with X-ray Guinier diffractograms at 1 bar with CuKα1 radiation. The results are summarized in Table 7. In agreement with the experiments of Ehrenberg et al. [19], the phase transition occurs between 200 and 100 K, and also the lattice parameter at 300 K $a = 6.7970(3)$ Å (Table 7) by X-ray diffraction is in good agreement with those of Ehrenberg by neutron diffraction [6.7905(1) and 6.8014(3) Å, averaged to 6.7960 Å] [19]. At 10 K, the X-ray data are $a = 4.7704(2)$, $c = 6.7709(4)$ Å (Table 7) are also in good agreement with the averaged neutron data at 1.5 and 2 K ($a = 4.7712$, $c = 6.7835$ Å) [19], neglecting a temperature difference of approximately 8 K.

The deviation of the tetragonal lattice parameters of LiIn from the ideal ratio $c/a = \sqrt{2}$ lies between 0.30% and 0.36% (Table 7). The splitting of the cubic 620 reflection (NaTl-type structure, high temperature phase) into the three tetragonal reflections 116, 332, and 420 (tetragonal distorted NaTl-type structure, low temperature phase) is shown in Fig. 11.

In Table 8 space group, calculated total energy E^o, volume V_o (Å^3) for one f.u., and bulk modulus B_o (GPa) of LiIn, with the tetragonal distorted NaTl- and the tetragonal distorted β-brass-type structure (CuAu-type structure) at zero pressure and temperature, are summarized.

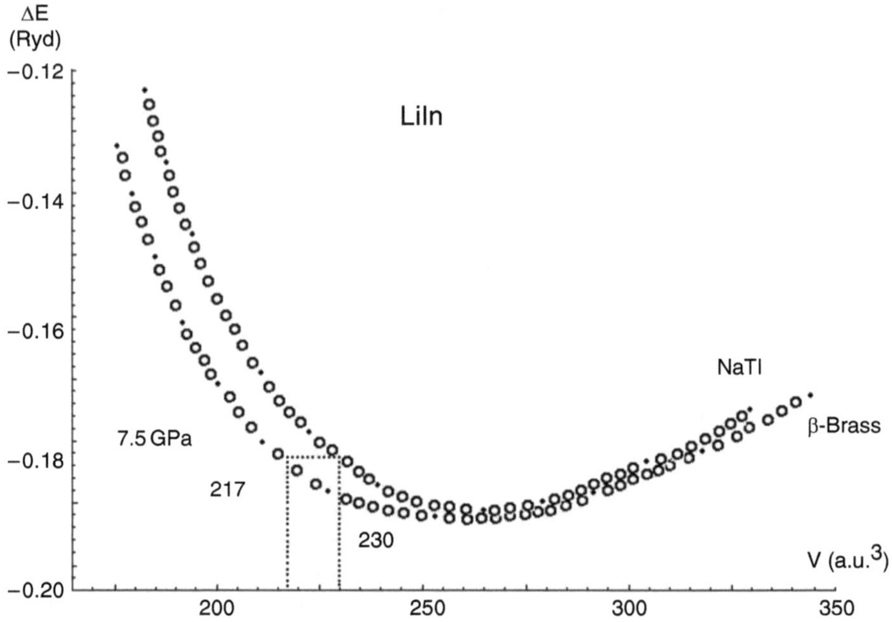

Fig. 10 Total energy E^o (Ryd) curves for LiIn with the NaTl- and the β-brass-type as function of the unit-cell volume V (a.u.^{3}). At 7.7 GPa, the NaTl-type structure of LiIn has a volume of 230 a.u.^{3}, and the β-brass-type structure one of 217 a.u.^{3}. From these data, the lattice parameters for both phases are derived (see text)

Comparison of Tables 6 and 8 shows that for LiIn the tetragonal distorted NaTl-type structure has 0.07 Å^3 lower calculated volume (= 0.2%) than the cubic phase. The total energy difference is zero, and the bulk moduli are practically equal (36, 37 GPa, respectively, Tables 6 and 8). The tetragonal distorted β-brass-type structure for LiIn has a 0.18 Å^3 lower volume (= 0.5%) than the cubic phase. The total energy difference for the two β-brass arrangements for LiIn is with 0.1 mRyd very low, and the bulk moduli are 35 GPa for the cubic phase and 39 GPa for the tetragonal distorted phase (Tables 6 and 8).

The energy differences between the NaTl- and the β-brass type for NaIn, KIn, RbIn, and CsIn are much higher than those for LiIn. They increase from –16.8, –29.7, –29.8 and –26.0 mRyd, respectively (Table 8), in comparison to +1.1 mRyd for LiIn. Therefore, there it seems to be difficult to transform NaIn, KIn, RbIn, and CsIn with the NaTl-type structure into the β-brass-type structure in the pressure range up to 30 GPa.

The estimated energies of formation $\Delta E^o_{f,AB}$ at zero temperature and pressure of NaTl-type phases for LiIn to CsIn and the estimated pressure (GPa) for stabilizing the NaTl-type structure in KIn to CsIn are summarized in Table 9.

For 1:1 alkali indides, LiIn and NaIn are the only stable phases with the NaTl-type structure at 1 bar. The estimated energies of formation (WIEN2k) at zero pressure and temperature $\Delta E^o_{f,0K,0GPa}$ are –51 and –18 kJ/mole, respectively.

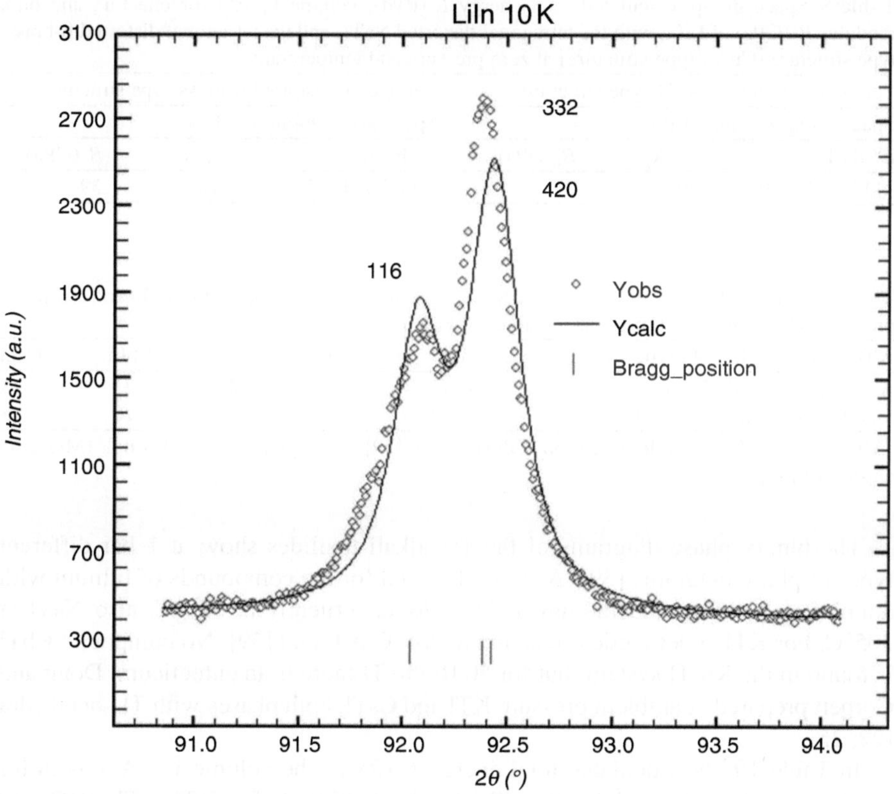

Fig. 11 Section of the Guinier diffractogram (CuKα1 radiation) at 10 K showing the splitting of the cubic 620 reflection into three tetragonal reflections 116, 332, and 420

With increasing metal radius from K to Cs, the estimated endergonic formation energies increase (+4, +21, +26 kJ/mole for KIn, RbIn, and CsIn, respectively; Table 8). There is good agreement found for the energy of formation $\Delta E^{o}_{f,AB,0K,0GPa}$(kJ/mole) for LiIn. By WIEN2k, a value of –51 kJ/mole was calculated. Predel's data are –49 kJ/mole (solution calorimetry) [29] and –46 kJ/mole (Miedema approach) [29, 30] (compare Sect. 2). The estimated stabilization pressure for the alkali indides KIn to CsIn with the NaTl-type structure lies in the range between 5 and 10 GPa (Table 9).

5.4 *Thallides LiTl, NaTl, KTl, RbTl, and CsTl*

As in the cases of aluminides, gallides, and indides, in this chapter the total energy E^o calculations for thallides are presented, but in addition the results of high pressure in situ investigations in DAC on the thallides up to 30 GPa are given.

Table 8 Space group, calculated total energy E^o(Ryd), volume V_o(Å^3) for one f.u., and bulk modulus B_o(GPa) of LiIn, with the tetragonal distorted NaTl- and the tetragonal distorted β-brass-type structure (CuAu-type structure) at zero pressure and temperature

Tetragonal distorted NaTl-type structure			Tetragonal distorted β-brass-type structure		
Space group $I4_1$/amd, LiIn			Space group P4/mmm, LiIn		
E^o(Ryd)	V_o(Å^3)	B_o(GPa)	E^o(Ryd)	V_o(Å^3)	B_o(GPa)
−11,780.19109	39.39	37	−11,780.19226	37.51	39

Table 9 Estimated energy of formation $\Delta E^o{}_{AB,f,0K,0GPa}$ (kJ/mole) for LiIn to CsIn and estimated pressure (GPa) for NaTl-type phases stabilized in KIn to CsIn

$A^I B^{III}$ phase with NaTl-type structure	LiIn	NaIn	KIn	RbIn	CsIn
$\Delta E^o{}_{AB,f,0K,0GPa}$ (kJ/mole)	−51[a]	−18	+4	+21	+26
Stabilization pressure (GPa)	0	0	5	10	5

[a] $\Delta H^\circ{}_{f,800K} = -49$ kJ/mole (solution calorimetry) [29], $\Delta H^\circ{}_{f,300K} = -46$ kJ/mol (Miedema approach) [30]

The binary phase diagrams of the 1:1 alkali thallides show at 1 bar different types of phase diagrams [39]. As also observed for the compounds of lithium with aluminum, gallium, and indium, LiTl melts congruently at 520°C, also NaTl at 305°C. For KTl, a peritectical reaction at 268°C is found [39]. No compound RbTl is found in the Rb-Tl system, but for 50 Rb:50 Tl there is an eutecticum. Dong and Corbett prepared at ambient pressure KTl and CsTl, both phases with Tl_6-octahedra [22, 25].

In Table 10, the calculated total energy E^o(Ryd), the volume V_o (Å^3), both for one f.u., and the bulk modulus B_o (GPa) for the thallides LiTl, NaTl, KTl, RbTl, and CsTl for the NaTl- and for the β-brass-type structure at zero pressure and temperature are summarized.

For the compounds LiTl and NaTl, both stable at ambient pressure, one calculates from the data of Table 10 the lattice parameters $a_{0K,0GPa} = 3.43$ Å for LiTl with the β-brass-type structure and $a_{0K,0GPa} = 7.50$ Å for NaTl with the NaTl-type structure. The tabulated lattice parameter [32, 33] for LiTl is $a_{300K,1bar} = 3.431$ Å and for NaTl it is $a_{300K,1bar} = 7.469$ [32], 7.462, 7.488, 7.448 [33] Å, with an average value of 7.467 Å leading to unit-cells with a volume of 40.39 and 416.3 Å^3, respectively. After correction for thermal expansion, $a_{10K,1bar} = 3.41$ Å (LiTl) and $a_{10K,1bar} = 7.42$ Å (NaTl) are obtained. The calculated lattice parameters of Schmidt are $a_{0K,0GPa} = 3.425$ Å (LiTl) and $a_{0K,0GPa} = 7.488$ Å (NaTl) [9]. From Christensen [18], only the lattice parameter of NaTl ($a_{0K,0GPa} = 7.439$ Å) can be compared here, because for LiTl an erroneous value of $a_{0K,0GPa} = 6.883$ Å is listed in Table 3. This value must belong to a not stable NaTl-type structure for LiTl.

As it has been also observed for aluminides, gallides, and indides, the differences in energy ΔE^o and volume ΔV_o between the phases with NaTl- and β-brass-type structure (Table 10) are strongly size dependent, having in mind that LiTl with β-brass-type structure is −10.6 mRyd more stable than NaTl-type phase.

Table 10 Calculated total energy E^o(Ryd), volume V_o(Å^3), and bulk moduli B_o (GPa) for one f.u. of thallides LiTl, NaTl, KTl, RbTl, and CsTl with the NaTl- and the β-brass-type structure at zero pressure and temperature

	NaTl-type structure			β-Brass-type structure		
	E^o(Ryd)	V_o(Å^3)	B_o(GPa)	E^o(Ryd)	V_o(Å^3)	B_o(GPa)
LiTl	–40,590.24904	43.24	31	–40,590.25935	40.20	32
NaTl	–40,899.78644	52.67	24	–40,899.78043	53.79	22
KTl	–41,778.89178	65.25	15	–41,778.87256	72.57	13
RbTl	–46,537.11216	71.23	14	–46,537.09218	80.02	13
CsTl	–56,154.18300	74.35	19	–56,154.15965	83.59	14

Christensen calculated this energy difference to −8.6 mRyd [18], and Schmidt with the scalar relativistic augmented-plane-wave method (SRAPW) to −11.0 mRyd [9]. From NaTl to CsTl, the differences in energy and volume increase (Table 10). For NaTl, the energy difference is $\Delta E^o = -6.0$ mRyd and the difference in volume $\Delta V_o = -1.12$ Å^3; for CsTl these values are −23.3 mRyd and −9.24 Å^3, respectively. For NaTl, this energy difference was calculated by Christensen to −8.8 mRyd [18] and by Schmidt to −0.4 mRyd [9].

Inspection of Table 10 shows also that LiTl with the NaTl-type and the β-brass-type structure has the highest bulk moduli B_o (31, 32 GPa, respectively) of the thallides series. Christensen calculated the bulk modulus B_o of LiTl with the β-brass-type structure to 35.7 GPa [18] and for NaTl with the NaTl-type structure to 29.9 GPa. The LiTl phases are therefore the hardest materials for the alkali thallides. Interestingly, LiTl is the only exception for the A^IB^{III} phases, calculated here with WIEN2k, where the β-brass-type phase has a higher bulk modulus B_o (32 GPa) than the NaTl-type phase (31 GPa, Table 10). Again the bulk modulus of CsTl (NaTl-type structure) is higher than for RbTl as it is also observed for CsIn (NaTl-type structure) in comparison to RbIn (Table 6).

In Table 11, the estimated energies of formation $\Delta E^o_{AB,f,0K,0GPa}$ (kJ/mole) at zero temperature and pressure and the estimated stabilization pressure to obtain the NaTl-type structure are summarized. For 1:1 alkali thallides, there are only two stable phases, 1 bar: LiTl with the β-brass- and NaTl with NaTl-type structure. The estimated energies of formation (WIEN2k) at zero pressure and temperature $\Delta E^o_{f,0K,0GPa}$ are –42 and –10 kJ/mole, respectively (Table 11). There is good agreement for the energy of formation $\Delta E^o_{f,AB,0K,0GPa}$ (kJ/mole) for LiTl. By WIEN2k, a value of –42 kJ/mole was obtained, Predel's data are −40 kJ/mole (solution calorimetry) [29] and –55 kJ/mole (Miedema approach) [29, 30]. With increasing metal radius from K to Cs, the endergonic formation energies increase (+4, +19, +23 kJ/mole for KTl, RbTl, and CsTl with the NaTl-type structure, respectively). The estimated stabilization pressure for these NaTl-type phases is about 5 GPa (Table 11). This is in good agreement with the facts that in the DAC experiments these phases can be obtained with the NaTl-type structure at "low" high pressures.

Table 11 Estimated energy of formation $\Delta E^{o}_{f,AB,0K,0GPa}$ (kJ/mole) for LiTl with the β-brass-, for NaTl, KTl, RbTl, and CsTl with the NaTl-type structure and estimated pressure (GPa) for NaTl-type phases stabilized in KTl to CsTl

$A^{I}B^{III}$ phase with NaTl-type structure	LiTl[a]	NaTl	KTl	RbTl	CsTl
$\Delta E^{o}_{f,AB,0K,0GPa}$ (kJ/mole)	−42[b]	−10	+4	+19	+23
Stabilization pressure (GPa)	0	0	5	5	5

[a]LiTl with β-brass-type structure
[b]$\Delta H^{o}_{f,800K}$ = –40 kJ/mole (solution calorimetry) [29], $\Delta H^{o}_{f,300K}$ = –55 kJ/mol (Miedema approach) [30]

Figure 12 shows that the β-brass-type phase for LiTl is lower in energy and volume. Therefore, at high pressure the β-brass type cannot be transformed in the investigated high pressure range up to 40 GPa into the NaTl-type structure.

The investigations in the DAC showed that the cubic β-brass-type structure of LiTl can be transformed into the tetragonal AuCu-type structure. The WIEN2k calculations showed that the equilibrium volume V_o for the AuCu phase (40.29 Å^3) is 0.09 Å^3 slightly higher than that of the cubic β-brass-type phase of LiTl (V_o = 40.20 Å^3). The difference in total energies is with 0.1 mRyd also very small. The β-brass-type phase is –0.6 mRyd more stable than the AuCu phase, having in mind that the NaTl-type phase of LiTl is with +10.3 mRyd less stable than the cubic β-brass-type phase (Table 10). The differences in energy and volume between the cubic β-brass-type structure and its tetragonal distortion are too small to be displayed with the scale of Fig. 12.

The DAC experiments on LiTl showed that at pressures up to 31.0(10) GPa both, the β-brass-type structure and the AuCu-type structure, are obtained. Figure 13 shows the Rietveld plot obtained at 21.0(8) GPa, analyzed with FullProf [28]. The lattice parameter for LiTl with the cubic β-brass- and the tetragonal AuCu-type structure up to 31.0(10) GPa is summarized in Table 12. Inspection of this table shows that the tetragonal phase has a slightly smaller unit-cell volume than the cubic one, e.g., 26.3(5) and 27.8(3) Å^3, respectively, at 31.0(10) GPa (Table 12). The averaged axial ratio *a*/*c* is 1.48. Calculation of the axial ratio at constant volume with WIEN2k resulted in a ratio *a*/*c* of 1.46.

The Rietveld analysis of the AuCu-type structure of LiTl at high pressure showed that 30% of the Tl atoms are distributed on Li positions. A lot of compression and decompression cycles of the LiTl sample in the DAC did not improve the ordering of the Li and the Tl atoms in the CuAu-type structure phase.

Interestingly, at ambient pressure Zintl, Brauer, and Dullenkopf determined a tetragonal distorted CsCl structure (β-brass-type structure) for the $A^{I}B^{V}$ compounds LiBi (a_{CsCl} = 3.361, c = 4.247 Å) [37] and for NaBi (a_{CsCl} = 3.46, c = 4.80 Å) [38] with one f.u./cell (z = 1). These lattice parameters can be transformed into that of the AuCu-type structure with z = 2. This results into the lattice parameters for LiBi with a = 4.753, c = 4.247 Å (a/c = 1.12) and for NaBi with a = 4.89, c = 4.80 Å (a/c = 1.02) (Table 13). These lattice parameters for LiBi and NaBi can now be compared with the calculated lattice parameters for the AuCu-type structures in $A^{I}B^{III}$ compounds LiTl and NaTl.

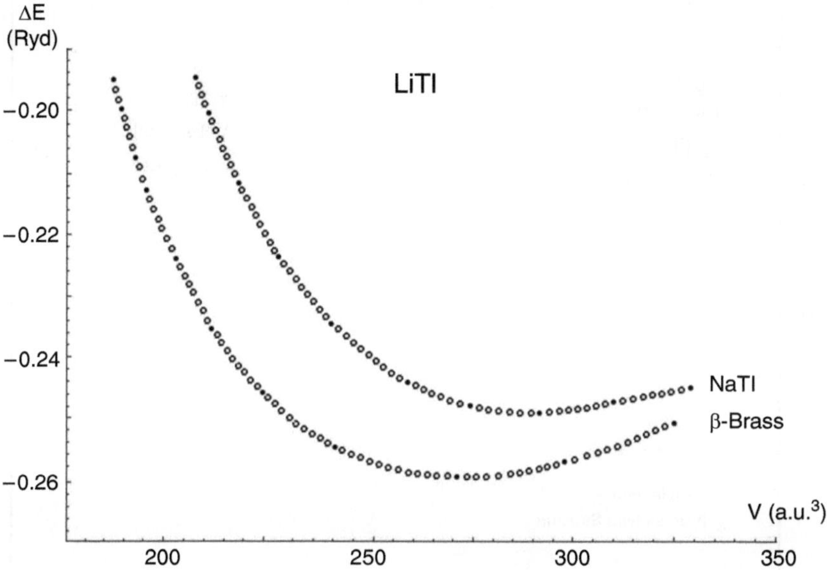

Fig. 12 Total energy E^o (Ryd) curves for LiTl with the β-brass- and the NaTl-type as function of the unit-cell volume V (a.u.^{3}). The β-brass type is lower in energy and in volume and cannot be transformed into the NaTl-type structure

The *a*/*c* ratio for LiTl and NaTl is with 1.46 remarkably larger than that for LiBi and NaBi with 1.12 and 1.02, respectively (Table 13). The lattice parameters for axis *a* (Å) have comparable size (4.753–5.38), but the lattice parameters for axis *c* the thallides are approximately 1 Å smaller than the bismuthides. The smaller *c*-axis in LiTl and NaTl leads to eight smaller A-B contacts in the thallides (2.97, 3.26 Å) and then in the bismuthides (3.19, 3.43 Å).

It is useful to compare now the A-B distances in the two phases of LiTl, one with the AuCu- and the other with the β-brass-type structure at 31.0(10) GPa. For the AuCu-type structure, an ordered distribution of Li and Tl atoms is assumed. With the lattice parameter 3.03(1) Å for the β-brass-type phase (Table 12), eight LiTl distances of 2.62(1) Å could be achieved. For the AuCu-type structure with the lattice parameters of 4.28(2) and 2.87(2) Å, eight LiTl distances of 2.58(1) Å could be realized under ideal conditions of ordered distribution of atoms. At 31.0(10) GPa, the LiTl distance in the AuCu-type structure, the eight LiTl distances are with 0.04 Å slightly smaller than in the β-brass-type structure. Therefore, this structure could be the end member obtained by very high pressure treatment of A^IB^{III} compounds with small radius ratio r_A/r_B, e.g., for lithium and the sodium compounds of this series. For larger A^I metals like potassium, rubidium, and cesium, the NaTl-type structure is favored as the DAC investigations for NaTl, KTl, and RbTl show in the following.

The difference between the NaTl-type and β-brass structure in NaTl is only –6.0 mRyd with the first also lower in volume (Table 10). With increasing pressure the

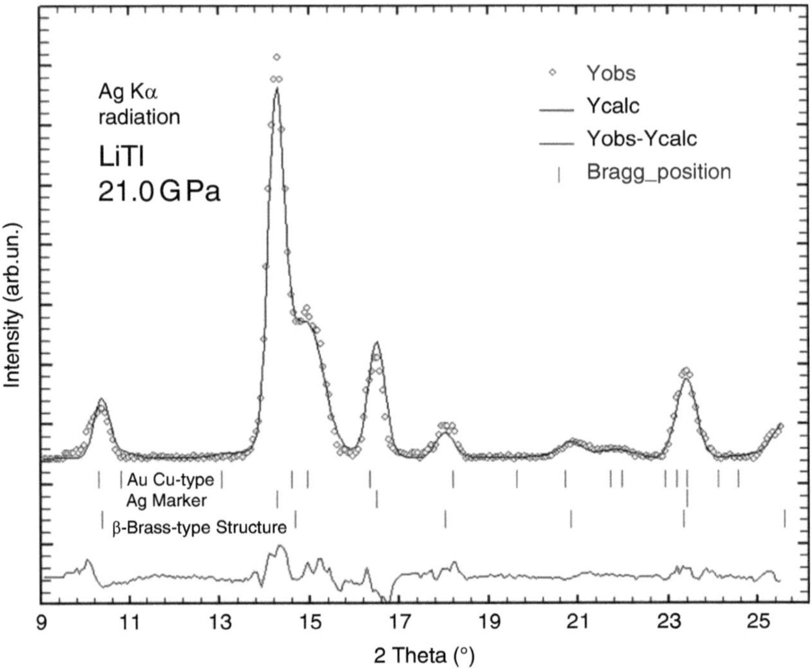

Fig. 13 Rietveld plot with FullProf [28] for LiTl with the β-brass- and the AuCu-type structure at 21.0 GPa (DAC diffractometer, Ag Kα radiation with 0.5609 Å). The measured data are shown as *circles*, the calculated diffractogram is shown as *dark*, the difference as *gray line*. The calculated 2-theta values are shown as *vertical bars*. Silver was used as internal pressure marker. The lattice parameter of silver at ambient pressure is $a = 4.0853$ Å, at the applied pressure in the DAC it is $a = 3.905$ Å. With $d/d^o = 3.905/4.0853 = 0.9559$, the applied pressure is 21.0(8) GPa [27]

Table 12 Lattice parameters of LiTl with the β-brass- and the AuCu-type structure to 31.0 GPa

p (GPa)	β-Brass-type structure		AuCu-type structure			
LiTl	*a*(Å)	*V* (Å^3)	*a*(Å)	*c*(Å)	*a*/*c*	*V* (Å^3)
0.0001	3.4332(5)[a]	40.47(1)				
5.2(3)	3.26(1)	34.6(3)				
11.0(5)	3.20(1)	32.8(3)	4.54(2)	3.07(2)	1.48	31.6(5)
12.8(5)	3.17(1)	31.9(3)	4.53(2)	3.05(2)	1.49	31.3(5)
14.0(6)	3.16(1)	31.6(3)	4.49(2)	3.03(2)	1.48	30.5(5)
21.0(8)	3.10(1)	29.8(3)	4.40(2)	2.98(2)	1.48	28.8(5)
24.5(8)	3.07(1)	28.9(3)	4.36(2)	2.94(2)	1.48	27.9(5)
31.0(10)	3.03(1)	27.8(3)	4.28(2)	2.87(2)	1.49	26.3(5)

[a]1 bar, Guinier diffractogram

difference in energy and volume decrease, as it is shown in Fig. 14. Here the total energy E^o is plotted as function of the volume. At about 210 a.u.^{3} (= 31.12 Å^3), the energy curves touch each other at about 40 GPa. At this volume both phases should be in equilibrium.

Table 13 Tetragonal lattice parameters *a* and *c*, axial ratio *a*/*c*, and volume *V* (*z* = 1) for four phases with the AuCu-type structure: LiBi and NaBi at ambient conditions [37, 38], LiTl and NaTl calculated with WIEN2k at zero pressure and temperature

	LiTl	LiBi	NaTl	NaBi
a(Å)	4.90	4.753	5.38	4.89
c(Å)	3.35	4.247	3.68	4.80
a/*c*	1.46	1.12	1.46	1.02
V (Å^3)	40.29	47.97	53.25	57.39
8 A-B (Å)	2.97	3.19	3.26	3.43

In Fig. 15, the Rietveld plot for NaTl with the NaTl- and the β-brass-type structure at 25.5 GPa is shown.

In Table 14, the lattice parameters of NaTl with the NaTl- and with the β-brass-type structure up to 32.5 GPa are summarized.

Inspection of Table 14 shows the lattice parameters of NaTl with the NaTl- and with the β-brass-type structure up to 32.5 GPa. The novel β-brass-type phase for NaTl created at high pressure has a statistical distribution of Na and Tl atoms. Annealing did not improve the ordering. NaTl, although not ordered, is after LiTl and LiIn the third A^IB^{III} phase with β-brass-type structure.

In Fig. 16, the total energy E^o curves for KTl with the orthorhombic ambient pressure phase with isolated Tl_6-octahedra [22], the NaTl- and the β-brass-type structure are shown.

In the earlier HP investigation on KTl in our group [21], a Rietveld plot obtained at 4.5 GPa had been published. In this investigation, the pressure has been largely increased. In Fig. 17, the Rietveld plot for the diffractogram of KTl obtained in the DAC diffractometer at 24.5 GPa is displayed, and in Table 15 the lattice parameter and volume (Å^3) for $z = 1$ are summarized. No β-brass-type phase could be obtained for KTl up to 30 GPa since here the differences in energy between both structures $\Delta E^o = -19.2$ mRyd (Table 10) are too large. The NaTl-type structure of KTl could be stored at ambient pressure after total pressure release. Diffractograms at 1 bar indicated a mixture of the cubic and the orthorhombic phase. The cubic phase remained for few days, and after this time only the orthorhombic phase was observed.

For KTl with the NaTl-type structure and also with the β-brass-type structure, electron density maps have been calculated with WIEN2k, both at 0 and 30 GPa. The contour line shows the calculated electron densities with increasing electron density from 0.05 to 0.95 in e/Å^3 (Figs. 18 and 19).

In Fig. 18, views are displayed on the (110) plane of KTl with the NaTl-type structure with zigzag chains of four Tl atoms with four adjacent K atoms. At 0 GPa (Fig. 18a), a contour line of 0.1 e/Å^3 surrounds the four Tl atoms indicating covalent bonding between them. The Tl–Tl bond distance in KTl at 0 GPa is 3.47 Å, and the unit-cell volume 514 Å^3. Between the Tl and the K atoms, there are kidney-shaped areas with a contour line of 0.05 e/Å^3 probably resulting from polarizing effects of the K ions on the valence electrons of the Tl polyanion [20].

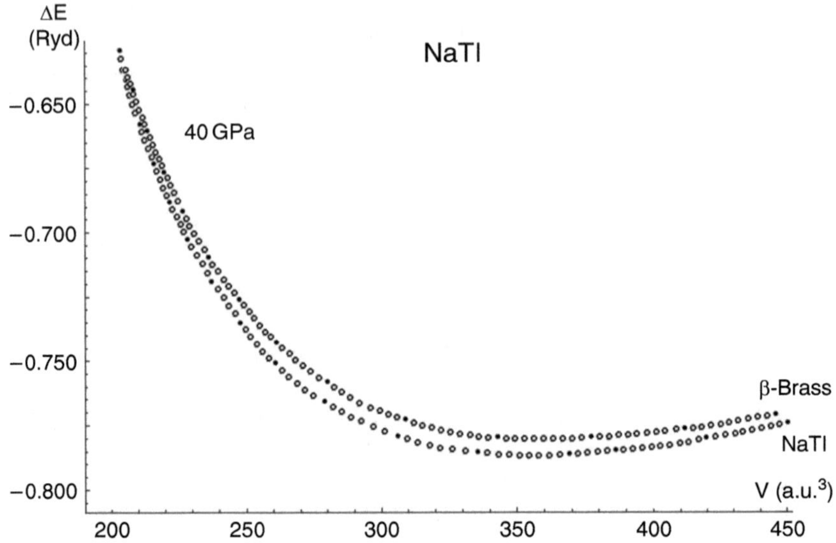

Fig. 14 Total energy E^{o} (Ryd) curves for NaTl with the NaTl- and the β-brass as function of the unit-cell volume V (a.u.^{3}). At approximately 210 a.u.^{3}, the NaTl- and the β-brass-type structure should be in equilibrium

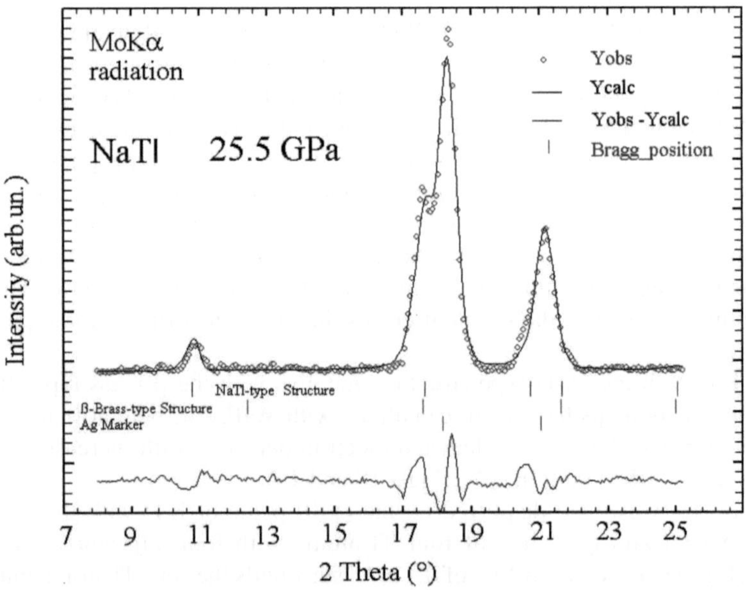

Fig. 15 Rietveld plot with FullProf [28] for NaTl with the NaTl- and the β-brass-type structure at 25.5 GPa (DAC diffractometer, MoKα radiation wit 0.7107 Å). The measured data are shown as *circles*, the calculated diffractogram is shown as *dark*, and the difference as *gray line*. The calculated 2-theta values are shown as *vertical bars*. Silver was used as internal pressure marker

Table 14 Lattice parameter of NaTl with the NaTl- and the β-brass-type structure up to 32.5 GPa, obtained by Rietveld analysis of the DAC diffractograms

p (GPa)	NaTl-type structure		β-Brass-type structure	
NaTl	*a*(Å)	*V* (Å^3)	*a*(Å)	*V* (Å^3)
0.001	7.482(2)[a]	52.36(4)		
0.5(2)	7.47(2)	51.1(3)		
3.2(3)	7.20(2)	46.7(3)	3.64(1)	48.2(1)
6.5(3)	7.06(2)	44.0(3)	3.52(1)	43.6(1)
16.0(6)	6.68(2)	37.3(3)	3.35(1)	37.6(1)
18.0(6)	6.66(2)	36.9(3)	3.32(1)	36.6(1)
22.0(8)	6.59(2)	35.8 (3)	3.29(1)	35.6(1)
25.5(8)	6.56(2)	35.3(3)	3.28(1)	35.3(1)
32.5(10)	6.43(2)	33.2(3)	3.20(1)	32.8(1)

[a]1 bar, Guinier diffractogram

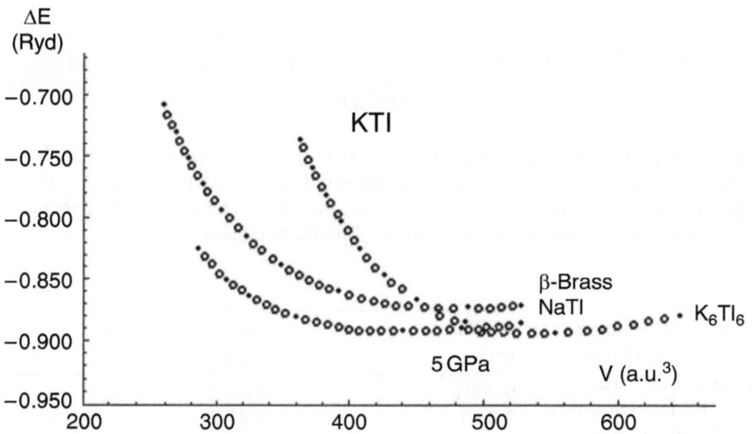

Fig. 16 Total energy E^o (Ryd) curves for KTl with the NaTl-, the β-brass- and the orthorhombic normal pressure phase, which contains isolated Tl_6-octahedra [22], are shown as function of the volume V (a.u.^{3}). At 480 a.u.^{3} (= 71.12 Å^3), the curve of the orthorhombic phase intersects the curve for NaTl-type phase. The NaTl-type structure for KTl is then lower in energy and volume. From the Murnaghan fit [34] for the orthorhombic phase, a pressure of 5 GPa is derived

At 30 GPa (Fig. 18b), KTl with the NaTl-type structure is compressed (−40%) to a unit-cell volume of only 306 Å^3 with a Tl–Tl bond distance of 2.92 Å^3. The 32 valence electrons in the unit-cell are now distributed into a smaller unit-cell. Now three contour lines of 0.1, 0.25, and 0.30 e/Å^3 surround the zigzag chain of four Tl atoms, indicating that even at 30 GPa KTl is even a Zintl compound K^+Tl^- with four covalent Tl–Tl bonds. In comparison to Fig. 18a, the kidney-shaped areas between the Tl and the K atoms are now increased in Fig. 18b and show a contour line of 0.15 e/Å^3, indicating an increased polarizing effect of the K ions.

In Fig. 19, views are displayed on the (110) plane of KTl with the β-brass-type structure. KTl with this structure is endergonic up to 30 GPa and could therefore not

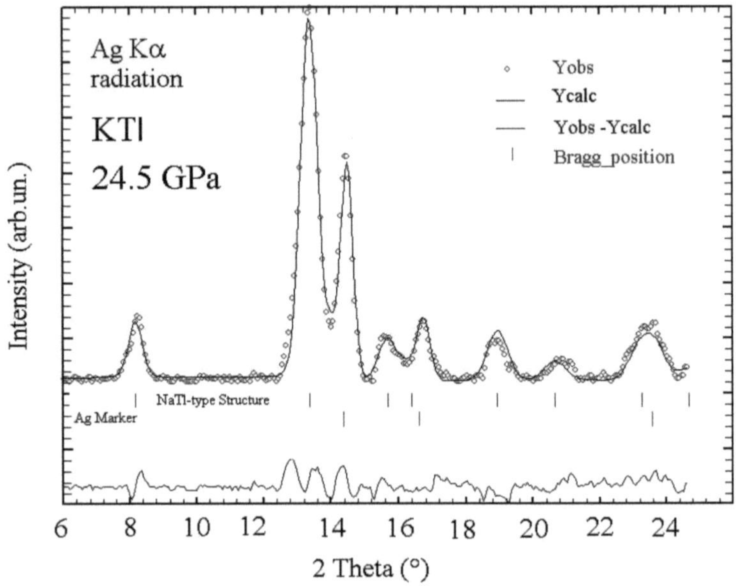

Fig. 17 Rietveld plot with [28] for KTl with the NaTl-type structure at 24.5 GPa (DAC diffractometer, Ag Kα radiation with 0.5609 Å). The measured data are shown as *circles*, the calculated diffractogram is shown as *dark*, the difference as *gray line*. The calculated 2-theta values are shown as *vertical bars*. Silver was used as internal pressure marker

Table 15 Lattice parameter of KTl with the NaTl-type structure up to 24.5 GPa, obtained by Rietveld analysis of the DAC diffractograms. The lattice parameter at 1 bar (0.0001 GPa) was obtained from a mixture with the orthorhombic phase [22] when the pressure was completely released

p (GPa)	NaTl-type structure	
KTl	*a*(Å)	*V* (Å^3)
0.0001	8.01(2)	64.2(5)
2.4(3)	7.81(2)	59.5(5)
3.0(3)	7.69(2)	56.8(5)
3.7(3)	7.63(2)	55.5(5)
7.2(4)	7.43(2)	51.3(5)
8.0(4)	7.40(2)	50.7(5)
14.6(6)	7.08(2)	44.4(5)
24.5(8)	6.82(2)	39.7(5)

be prepared in a DAC at this pressure. At 0 GPa (Fig. 19a), the Tl and the K atoms show a nearly spherical charge distribution with contour lines increasing from 0.1, 0.15, and 0.2 e/Å^3 to much higher values. Between the Tl and the K atoms, there are areas found with a contour line of 0.1 e/Å^3. These areas do not touch neither the Tl nor the K atoms. The unit-cell volume is 70.8 Å^3. At 30 GPa (Fig. 19b), the unit-cell volume of KTl with the β-brass type is compressed (–40%) to 41.4 Å^3. The Tl and the K atoms show also a nearly spherical charge distribution, but now with contour lines increasing from 0.2, 0.25, and 0.3 e/Å^3 to much higher values. Between the

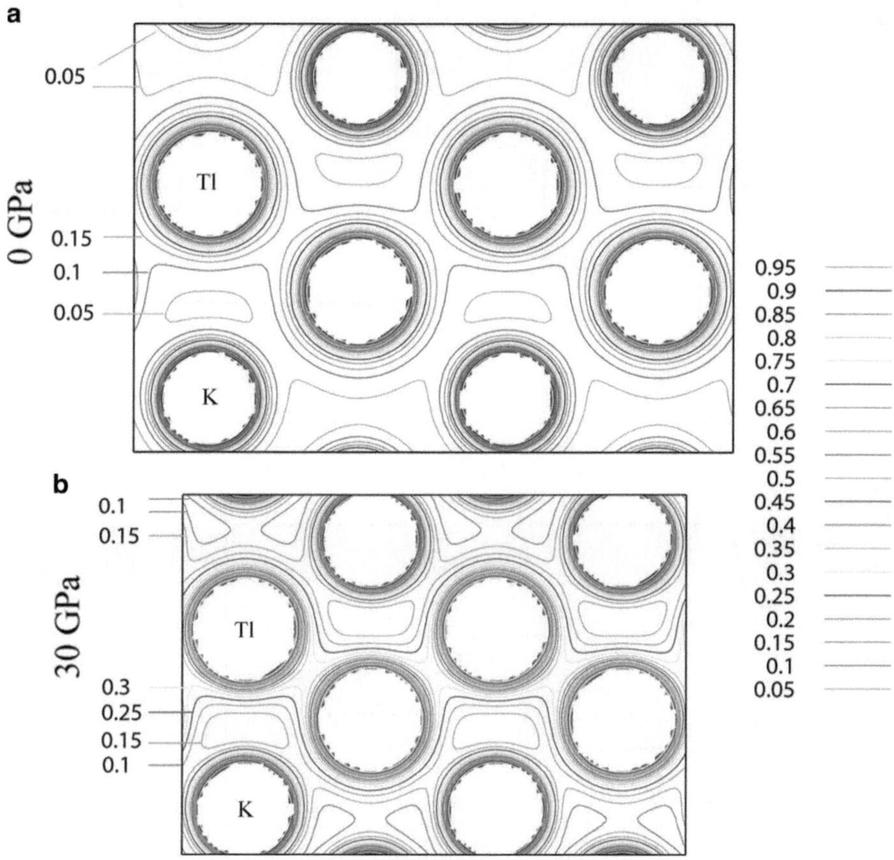

Fig. 18 Electron density maps for KTl with the NaTl-type structure [23] shown for the (110) plane with a zigzag chain of four Tl atoms and four adjacent K atoms. Isolines are in equal steps of ≈0.05 e/Å^{-3} increasing from 0.05 to 0.95 in e/Å^3 (**a**) at 0 GPa and (**b**) at 30 GPa

Tl and the K atoms, there are areas with contour lines of 0.1 e/Å^3. Interestingly, a second contour line of 0.15 e/Å^3, lying between these atoms, occurs now. These lines surround now the Tl and also the four K atoms, probably indicating weak metallic bonding between them.

In Table 16, for RbTl with NaTl-type structure, the lattice parameters a(Å) and the volume V (Å^3) for one formula up to 16.8(6) GPa are summarized. Fitting of the four experimental pressure–volume data of Table 16 with the Murnaghan equation [34] leads to a lattice parameter of a = 8.37(4) Å at 1 bar and 300 K. By correction of this value by thermal expansion (derived from the thermal expansion of LiIn), a value of $a_{10K,1bar}$ = 8.32 Å is derived. By calculation with WIEN2k (Table 10), $a_{0K,\,=\,GPa}$ = 8.29 Å is obtained.

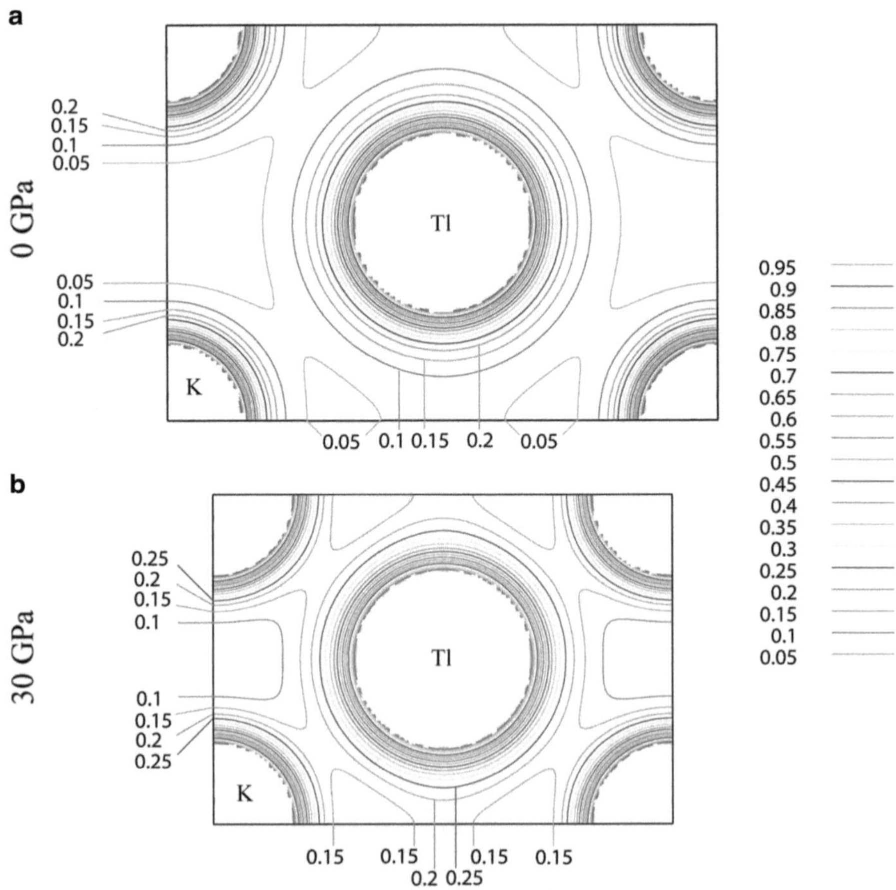

Fig. 19 Electron density maps for KTl with the β-brass-type structure [23] shown for the (110) plane. Isolines are in equal steps of ≈0.05 e/Å^{-3} increasing from 0.05 to 0.95 in e/Å^3 (**a**) at 0 GPa and (**b**) at 30 GPa

By the DAC experiments for CsTl, only diffractograms of low quality, showing strong texture, were obtained. Annealing at about 100°C did not improve their quality. Lattice parameters for CsTl with the NaTl-type structure could be derived indexing the strong reflection as 220. At 21 GPa, a lattice parameter of 7.23(8) Å and at 4 GPa one of 8.07(8) Å. From $V_o = 74.35$ Å^3 obtained with WIEN2k for CsTl (Table 10), a lattice parameter of $a_{0\text{K},0\text{GPa}} = 8.41$ Å is derived. Correction for thermal expansion leads to $a_{300\text{K},1\text{bar}} = 8.37$ Å for CsTl with the NaTl-type structure.

The estimations of the energy of formation for the $A^{I}B^{III}$ compound with WIEN2k have been internally standardized against the heats of formation obtained by solution calorimetry and its estimation by the Miedema approach of LiIn and LiTl, both from Predel's group. Six $A^{I}B^{III}$ of total 20 compounds are stable at 0 K,

Table 16 Lattice parameter of RbTl with the NaTl-type structure up to 16.8 GPa, obtained by Rietveld analysis of the DAC diffractograms. The parameter at 1 bar (0.0001 GPa) was calculated with WIEN2k and then corrected for thermal expansion with the data of LiIn

p (GPa)	NaTl-type structure	
RbTl	*a*(Å)	*V* (Å^3)
0.0001	8.37(4)[a]	73.3(10)
0.6(2)	8.01(2)	64.2(5)
7.6(3)	7.66(2)	56.2(5)
14.5(6)	7.48(2)	52.3(5)
16.8(6)	7.32(2)	49.0(5)

[a]8.32 Å calculated for 10 K, 0 bar, corrected for thermal expansion +10.38 Å^3 from LiIn-data +5.84 Å^3

0 GPa (LiAl, LiGa, LiIn, LiTl and NaIn, NaTl) in agreement with Zintl's classical investigations. The remaining 14 A^IB^{III} compounds are endergonic at zero pressure and temperature. The estimations show that in the pressure range up to 30 GPa, the NaTl-type structure could be stabilized. The lowest pressure (5 GPa) for stabilizing the NaTl-type structure is estimated for NaGa, KTl, RbTl, CsIn, and CsTl, and the highest pressure (30 GPa) for KAl, RbAl, and CsAl (Tables 3, 5, 9, and 11).

The lattice parameters calculated with WIEN2k (0 K, 0 GPa) deviate for LiAl, LiGa, LiIn, LiTl, NaIn, and NaTl from the experimental values (300 K 0.0001 GPa) after correction for thermal expansion only few on the second digit. For LiAl, the calculated lattice parameter with WIEN2k (0 K, 0 GPa) is 6.33 Å, and the experimental lattice parameter (10 K, 0.0001 GPa) is also 6.33 Å. For LiGa, LiIn, LiTl, NaIn, and NaTl, the analogous data in Å are 6.18/6.15, 6.81/6.75, 3.43/3.41, 7.30/7.29, and 7.50/7.42. Also the calculated lattice parameters of LiIn at 7.5 GPa, both for the NaTl- and the β-brass-type structure, are in very good agreement with the DAC experiments [20]. Therefore, there are now lattice parameters at 0 GPa available for 19 A^IB^{III} compounds with the NaTl-type structure with sufficient precision. LiTl with β-brass-type structure is more stable than that with the NaTl-type structure and is not considered here.

These data can be used to construct the near-neighbor diagram for the NaTl-type structure. Pearson had introduced such diagrams to get information which factors control intermetallic compounds with specific structures, e.g., a geometrical or a chemical bond factor [40–43]. Beside several intermetallic structures, e.g., the Laves phases, the σ-phases, and the CrB-phases, Pearson investigated also the NaTl-type structure for A^IB^{III} compounds and, in addition, for LiZn and LiCd. For the A^IB^{III} compounds, he corrected the radii for r_{AI} and r_{BIII} given by Teatum and Gschneidner [43–45] to coordination number eight and calculated with the known experimental lattice parameters of the NaTl-type phases the distance $d_A\left[d_A = (a/4)\sqrt{3}\right]$. Pearson used instead of the radii, the diameters D_A and D_B, where D_A is always taken as the greater one, and plotted then a strain parameter $(D_A - d_A)/D_B$ as function of diameter ratio D_A/D_B. He adjusted in the near-neighbor diagram also three lines calculated for 4 B-B, 4-4 A-B and 4 A-A contacts. In Pearson's original plot [40–43], there were seven data points obtained from the binary phases known at that time LiAl, LiGa, LiIn, NaIn, and NaTl and – as mentioned above – also LiZn and LiCd. Pearson's five data points for the A^IB^{III} phases are plotted in the left lower part of Fig. 20, not considering LiZn and LiCd. The three left lower points belong to "InLi" (In is here the greater A, Li the smaller B), LiAl and LiGa, and, the two slightly upper ones, to NaTl and NaIn. These five points lie above the line for "4 B-B" indicating that these contacts are compressed [46]. For the remaining 14 A^IB^{III} compounds, the lattice parameters with NaTl-type structure have been used, obtained at zero temperature and pressure by WIEN2k (without correction for thermal expansion). These data and the corrected radii r_{AI} and r_{BIII} [43, 44] were the input for Fig. 20. The following radii have been used: $r_{Li} = 1.509$, $r_{Na} = 1.858$, $r_K = 2.323$, $r_{Rb} = 2.493$, $r_{Cs} = 2.678$, $r_{Al} = 1.379$, $r_{Ga} = 1.358$, $r_{In} = 1.610$, $r_{Tl} = 1.663$, in Å [40, 43]. Such near-neighbor diagram is shown for such 19 A^IB^{III} compounds in Fig. 20.

Nearly 40 years ago, between 1968 and 1973, it was out of imagination for Pearson [40–43] that an A^IB^{III} compound of composition KTl with the NaTl-type structure could be prepared experimentally. At that time, high pressure experiments with a DAC were quite "exotic." Pearson concluded that "in KTl, however, the radius ratio (about 1.40), is so much larger, that compression of the K atom in order to achieve full Tl–Tl contacts is no larger profitable for formation of the NaTl arrangement" [42]. And also: ". . . even the most favorable case of KTl the value of $D_A/D_B = 1.397$ is too large for Tl–Tl contacts to be formed by compression of the K atoms" [47] and "potassium compounds do not have the NaTl structure since even for the largest group III element, thallium. . ." [41].

However, inspection of Fig. 20 shows that Pearson was only quite right at ambient pressure, but the data point for KTl with the NaTl-type structure lies only slightly below the line for 4 B-B contacts. This is an indication that the B-B contacts have to be stretched. In NaTl, itself, with a lattice parameter of $a = 7.467$ (10) Å at ambient pressure (Table 14), there is a Tl–Tl distance of 3.23(1) Å. On the other hand, in KTl with $a = 8.01(2)$ Å at ambient pressure (Table 15), the Tl–Tl distance is 3.47(1) Å. Therefore, the Tl–Tl distance in KTl has to be stretched by 7%.

Inspection of Fig. 20 shows also that the six data points from KTl to CsIn (with increasing D_A/D_B: KIn, RbIn, RbTl, CsTl) lie also on a line which could be obtained by extrapolation of Pearson's original five data points. However, this extrapolated line intersects the calculated line for 4 B-B contacts at approximately $D_A/D_B = 1.3$. Below $D_A/D_B = 1.3$, 4 A-A contacts have to be compressed, above this value 4 B-B contacts have to be stretched. It is evident from Fig. 20 that for the 19 A^IB^{III} compounds considered here, the smallest deviation to the calculated line for 4 B-B contacts is obtained for KTl to CsIn. Indeed, here the calculated pressure to obtain the NaTl-type structure for KTl and CsIn is only 5 GPa. The line for NaAl to NaGa has a medium deviation to the calculated line for 4 B-B contacts with stabilization pressures of 25 and 5 GPa. The largest deviation to the line for 4 B-B contacts is found for KAl to CsGa. Here the stabilization pressure is 30 and 20 GPa. This indicates that for these compounds the importance of 4 B-B contacts is lowered, but that of 4-4 A-B contacts is increased.

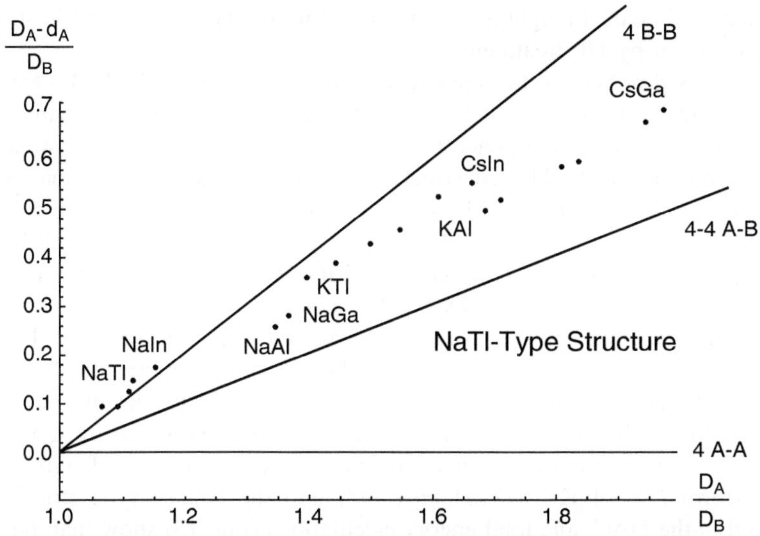

Fig. 20 Near-neighbor diagram for the NaTl-type structure at 1 bar. The data for the ratio of the atomic diameters at 1 bar were taken from Teatum et al. [43, 44] and corrected for coordination number eight [41]

From the total energy E^o curves, discussed in Sects. 5.1–5.4, it was derived that LiAl (D_A/D_B = 1.094), LiGa (1.111), InLi (1.067), and NaTl (1.117) (Fig. 20) could be suitable candidates for a high pressure transformation into the β-brass-type structure. For LiIn [20], the transformation into the β-brass-type structure was verified by DAC experiments in 1998. In this work, it was shown that for NaTl (1.117) (Fig. 14; Table 14) a β-brass phase could be obtained. For NaIn (1.154), the transformation pressure should be much higher. For LiAl (1.094) and LiGa (1.111), the HP transformations into the β-brass-type structure seem to be much easier.

6 Conclusion

The NaTl- and the β-brass-type structures in the 20 investigated A^IB^{III} compounds consist of interpenetrating cubes with 8:8 nearest neighbor coordination. Neglecting small differences, this is also true for the two tetragonal distortions. In the NaTl-type structure and its tetragonal distortion in space group $I4_1$/amd, 8:8 coordination is achieved for the A^IB^{III} compounds with 4 A-B contacts and, additionally, either with 4-4 A-A or 4-4 B-B contacts. In the ordered β-brass-type structure, the coordination consists of only 8:8 A-B contacts. The dimorphism of the NaTl- and the β-brass-type structures in a special compound depends on their total energies E^o. The WIEN2k calculations show that for small radius ratio $r_A{:}r_B$, the β-brass-type structure is favored, for large the NaTl-type structure. This ordering of the A^I and the B^{III} atoms in interpenetrating cubes with 8:8 nearest neighbor coordination for both typically metallic structures can be changed by HP treatment.

As discussed in Sect. 4, the site occupation is identical for the NaTl- and the β-brass-type structure, and also its two tetragonal distortions, neglecting small differences. There are two interpenetrating cubes which give 8:8 nearest neighbor coordination. In the NaTl-type structure and its tetragonal distortion in space group $I4_1/amd$, this is achieved by 4-4 B-B and 4 A-B contacts, if the cube is B-centered, or by 4-4 A-A and 4 A-B contacts, if the cube is A-centered. In the β-brass-type structure and its tetragonal distortion in space group P4/mmm, the 8:8 nearest coordination is achieved by 8:8 A-B contacts.

Pearson had stated that the "NaTl-type structure is neither ionic nor dominated mainly by strong diamond-type covalent bonds, but characteristically metallic phases adopting 8-8 coordination" [43]. However, electron density calculations for KTl with the NaTl-type structure (Fig. 18) clearly show that even at 30 GPa covalent bonding in the four-connected net of Tl atoms is present. Up to 24.5 GPa no transformation of KTl (D_A:D_B = 1.397) into the β-brass-type structure was observed in the DAC, and total energy calculations (Fig. 15) show that the energy difference between both structures is quite large.

For NaAl (1.347) and NaGa (1.368), the metallic radius of sodium seems also too large to obtain the β-brass-type structure in the same pressure range as above. The total energy E^o calculations with WIEN2k indicate that for larger D_A:D_B values (1.37–1.942) in comparison with β-brass-type structure, the NaTl-type structure is stabilized at pressures up to 30 GPa. The concurrence of the NaTl- and the β-brass-type structure is therefore strongly influenced by the ratio D_A:D_B.

For small D_A/D_B ratios (1.0–1.2), the transformation NaTl-type into the β-brass-type structure has been experimentally verified [20] in "InLi" (D_A:D_B = 1.067 at 1 bar) and in NaTl by this work (D_A:D_B = 1.117). Total energy E^o calculations show that it should be possible to perform this transformation in the pressure range up to 40 GPa also in LiAl (D_A:D_B = 1.094) and in LiGa (1.111). Therefore, the β-brass-type structures for LiAl, LiGa, and NaTl, which could be obtained by applying high pressure up to 40 GPa, simulate the behavior of LiTl at 1 bar. This structural change, obtained "between" NaTl and LiTl by only lowering the metal radius of Na (r_{Na} = 1.858 Å) to Li (r_{Li} = 1.509 Å), can be understood by applying "high internal pressure" at 1 bar.

Phases with low D_A:D_B ratios achieving packing of A^I and B^{III} spheres either in the β-brass- or in the AuCu-type structure are energetically favored over those with covalent B^{III}–B^{III} bonding in the NaTl-type structure. In addition, phases with an arrangement of A^I and B^{III} spheres should be more ductile (lower B_o) than those with covalent B^{III}–B^{III} bonding which should be more brittle (higher B_o).

Acknowledgments The author thanks Prof. Dr. Peter Kroll, Department of Chemistry and Biochemistry, The University of Texas at Arlington, for support in the first stages of the WIEN2k calculations on A^IB^{III} compounds, when it became evident that contrary to earlier published data, LiIn with the β-brass-type structure is lower in energy than the NaTl-type structure. Prof. Kroll confirmed this result by own calculations with the program VASP. The author also thanks Prof. Dr. Peter Blaha, Institut für Materialchemie, University of Vienna, for critical comments and many advices.

References

1. Zintl E, Woltersdorf G (1935) Z Elektrochem 41:876
2. Zintl E (1939) Angew Chem 1:52
3. Klemm W, Busmann E (1963) Z Anorg Allg Chem 319:297
4. Klemm W (1958) Proc Chem Soc London:329
5. Schäfer H, Eisenmann B, Müller W (1973) Angew Chem 85:742; Angew Chem Int Ed 12:694
6. Schäfer H (1985) Ann Rev Mater Sci 15:1
7. Hoffmann R (1988) Solids and surfaces: a chemist's view of bonding in extended structures. VCH Publishers, New York, p 3
8. Zintl E, Dullenkopf W (1932) Z Physik Chemie B 16:195
9. Schmidt PC (1985) Phys Rev B 31:5015
10. Corbett JD (2000) Angew Chem 112:682; Angew Chem Int Ed 39:670
11. Laves F (1941) Naturwissenschaften 29:244
12. Gottfried C, Schossberger F (1937) Strukturbericht, Band III, 1933-1935. Akademische Verlagsgesellschaft M.B.H, Leipzig, p 19
13. Zintl E, Neumayr S (1933) Z Physik Chem B 20:272
14. Zintl E, Brauer G (1933) Z Physik Chem B 20:245
15. Ewald PP, Hermann C (1931) Strukturbericht 1913–1928. Akademische Verlagsgesellschaft M.B.H, Leipzig, p 74
16. Maknovetskii AB, Krasko GL (1977) Phys Stat Sol B 80:341
17. Schmidt PC (1987) In: Clarke M, Chestnut H, Goodenough JB, Ibers JA, Evanston CK, Mingos DMP, Neilands JB, Palmer GA, Reinen D, Sadler PJ, Weiss R, Williams RJP (eds) Structure and bonding, solid state chemistry, vol 60. Springer Verlag, Berlin, p 91
18. Christensen NE (1985) Phys Rev B 32:207
19. Ehrenberg H, Pauly H, Hansen T, Jaud J-C, Fuess H (2002) J Solid State Chem 167:1
20. Schwarz U, Bräuninger S, Syassen K, Kniep R (1998) J Solid State Chem 137:104
21. Evers J, Oehlinger G (2000) Inorg Chem 39:628
22. Dong Z, Corbett JD (1993) J Am Chem Soc 115:11299
23. Blaha P, Schwarz K, Madsen G, Kvasnicka D, Luitz J (2001) WIEN2k: an augmented plane wave plus local orbital program for calculating crystal properties, November 2001. Vienna University of Technology, Vienna
24. Blase W, Cordier G, Müller V, Häussermann U, Nesper R, Somer M (1993) Z Naturforsch B 48:754
25. Dong Z-C, Corbett JD (1996) Inorg Chem 35:2301
26. Mao H, Bell PM (1978) Science 200:1145
27. Mao HK, Bell PM, Shaner JW, Steinberg DJ (1978) J Appl Phys 49:3276
28. Rodriguez-Carvajal JR (1990) FullProf: a program for Rietveld refinement and pattern matching analysis. In: Abstracts of the satellite meeting on powder diffraction of XV congress of the IUCr, Toulouse, France, 1990. International Union of Crystallography, Chester, p 127
29. Sommer F, Fischer B, Predel B (1982) In: Borgstedt HU (ed) Material behaviour and physical chemistry in liquid metal systems. Plenum Press, New York, p 395
30. Miedema AR, Chatel de PF (1979) Theory of alloy formation. In: Bennett LH (ed) Proceeding of a symposium of the AIME, annual meeting, New York 1980, pp 344–387
31. Wu Z, Cohen R (2006) Phys Rev B 73:235116
32. Villars P, Calvert LD (1985) Pearson's handbook for intermetallic phases. ASM International, Materials Park, First Printing 1985
33. Villars P, Calvert LD (1988) Pearson's handbook for intermetallic phases. ASM International, Materials Park, Second Printing, February 1996
34. Murnaghan FD (1944) Proc Nat Acad Sci U S A 30:244
35. Perdew JP, Wang Y (1992) Phys Rev B 45:13224
36. Perdew JP, Burke S, Ernzerhof M (1996) Phys Rev Lett 77:3865

37. Zintl E, Brauer G (1935) Z Elektrochem 41:297
38. Zintl E, Dullenkopf W (1932) Z Physik Chem B 16:183
39. Massalski TB (1990) Binary alloy phase diagrams, vol 1: Ac-Ag to Ca-Zn. ASM International, Materials Park, p 178
40. Pearson WB (1968) Acta Crystallogr B24:1415
41. Pearson WB (1969) In: Giessen BC (ed) Development in the structural chemistry of alloy phases. Plenum Press, New York, pp 45–48
42. McNeil MB, Pearson WB, Bennett LH, Watson RE (1973) J Phys C Solid State Phys 6:1
43. Pearson WB (1973) The crystal chemistry and physics of metals and alloys. Wiley-Interscience, New York, p 576
44. Teatum E, Gschneidner K, Waber J (1960) LA2345. US Dept of Commerce, Washington
45. Pearson WB (1973) The crystal chemistry and physics of metals and alloys. Wiley-Interscience, New York, p 151
46. Pearson WB (1973) The crystal chemistry and physics of metals and alloys. Wiley-Interscience, New York, p 53
47. Pearson WB (1973) The crystal chemistry and physics of metals and alloys. Wiley-Interscience, New York, p 578

Struct Bond (2011) 139: 97–142
DOI:10.1007/430_2010_25

Published online: 20 January 2011

Zintl Clathrates

Andrei V. Shevelkov and Kirill Kovnir

Abstract Zintl clathrates are the large family of inorganic compounds that combine aesthetically beautiful structures and interesting properties, including superconductivity and high thermoelectric efficiency. This review surveys various aspects of chemistry and physics of these compounds. It includes classification of Zintl clathrates, their crystal chemistry, electronic structure and properties, diverse synthetic routes to clathrates, as well as prospects of creation of new functional materials on their base. The applicability and limitations of the Zintl approach to the analysis of chemical bonding in clathrates are also contemplated.

Keywords Cage compounds, Clathrates, Crystal structure, Electronic structure, Superconductivity, Thermoelectric materials, Zintl phases

Contents

A.V. Shevelkov (✉)
Chemistry Department, Moscow Lomonosov State University, Moscow, Russia
e-mail: shev@inorg.chem.msu.ru

K. Kovnir
Department of Chemistry and Biochemistry, Florida State University, Tallahassee, FL, USA

1 Introduction

Chemistry is an experimental science. Students coming to study chemistry learn it with the help of experiments demonstrated, at first stages, by lecture assistants in lecture theaters. One of the amusing experiments they could watch in a demonstration is the apparent crystallization of ice upon passing chlorine gas through water cooled down to +7°C. Of course, what they see is not ice. It is a chlorine hydrate also known as the type I clathrate; its crystal structure is shown in Fig. 1. Surprisingly, many polar intermetallics and similar compounds adopt the same type of the crystal structure. In 1965, Kasper, Hagenmuller, Pouchard, and Cros reported the preparation of Na_8Si_{46} and Na_xSi_{136} by careful decomposition of NaSi and showed the remarkable similarity between the crystal structures of these compounds and some gas hydrates [1, 2]. After 45 years, the list of nonhydrate clathrates comprises more than 200 compounds having six different structure types and a great variety of the chemical composition. What combines them is that all these compounds (with just three exceptions) have a group 14 element as the main component of the four-bonded clathrate framework. Many of them are low-gap semiconductors, and the term "Semiconducting Clathrates" is frequently used. However, quite a number of these compounds display metallic properties; that is why the alternative name for this family is "Intermetallic Clathrates." This term also has its drawbacks, because a number of clathrates contain such elements as, for instance, phosphorus, tellurium, or bromine. Another option is to use the term "Inorganic Clathrates" to set them aside of ice hydrates. This term is also far from being perfect because the ice hydrates are in fact inorganic compounds by nature. Having no absolutely correct definition at hand, we will use the term "Zintl Clathrates" in this chapter since most of them are Zintl phases; however, even this name is not completely correct as there are compounds displaying metallic properties.

The term "Clathrate" originates from the Latin *clathratus* and means "protected by a cross bar." This term, introduced by Powell in 1948, refers to hydrates and

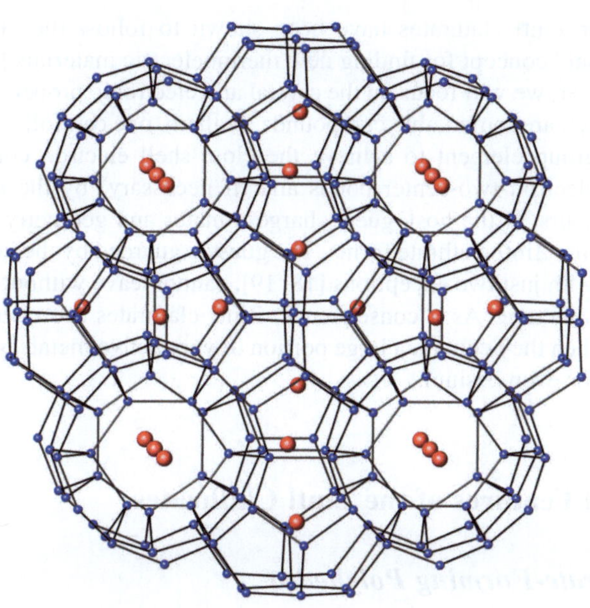

Fig. 1 Crystal structure of chlorine hydrate. Hydrogen atoms are omitted for clarity and the bare centers of chlorine molecules are schematically represented as *large spheres*

analogs with a three-dimensional structure of the host framework that traps guest atoms or molecules in its spacious cavities [3]. Although the history of the clathrates dates back to the beginning of the nineteenth century [4], it was only 1952 when Allen reported the crystal structure of the above-mentioned chlorine hydrate $Cl_2{\cdot}7.25H_2O$ [5]. In this compound, water molecules build up a three-dimensional framework with the help of hydrogen bonds, leaving two types of polyhedral cages for filling by chlorine molecules (Fig. 1). The chlorine molecules only partially fill the cavities, leading to the unit cell formula $6.34Cl_2{\cdot}46H_2O$ instead of the regular clathrate formula $8Cl_2{\cdot}46H_2O$ for the hydrate with the completely filled guest positions. By analogy, in Na_8Si_{46} all silicon atoms form the three-dimensional framework of the same topology and the sodium atoms occupy all guest positions [1]. Several types of the clathrate hydrate structure are known [6], but despite such diversity, the common feature of the clathrates is the inability of the host framework to survive in the absence of the guest.

For many years all clathrates were just curious compounds with fascinating structures. The situation had changed by the beginning of this millennium. Clathrate hydrates have been investigated as containers for transportation, storage, and even recovery of methane and carbon dioxide [7–10]. A lot of effort has been focused on preventing the ice clathrate formation in oil pipelines [10]. Recently, there were reports about the inclusion of molecular hydrogen into the hydrate framework [11, 12]. Zintl clathrates also show very interesting properties. Superconductivity was discovered in silicon- and germanium-based type II clathrates

[13–15]. Other Zintl clathrates have been shown to follow the "Phonon Glass, Electron Crystal" concept for finding new thermoelectric materials [16, 17].

In this chapter, we will focus on the crystal and electronic properties of the Zintl clathrates, which are remarkable compounds. In them, one can follow the tendency of any main-group element to achieve the close-shell electron configuration by forming two-electron two-center bonds and (if necessary) by the localization of lone electron pairs. If the host–guest charge transfer and geometry mismatch are adjusted, then the Zintl clathrate forms. The guest is guarded by the host framework and, as a rule with just two exceptions [18, 19], cannot leave without destruction of the clathrate structure. As a consequence, Zintl clathrates are chemically stable phases even when they contain a large portion of very active metals, such as barium or potassium, or even cesium.

2 General Features of the Zintl Clathrates

2.1 Clathrate-Forming Polyhedra

Atoms of the group 14 elements can form an infinite number of tetrahedrally bonded frameworks belonging to the class of the three-dimensional four-bonded networks [20]. The clathrate frameworks are distinguished by large polyhedral cages formed in response to the presence of guest atoms. These polyhedra have a great number of vertices, starting with 20. Each polyhedron is denoted by the number of five-member and six-member faces. For instance, the notation $[5^{12}6^2]$ indicates the polyhedron having 12 five-member faces and 2 six-member faces. All polyhedra known to form the structures of Zintl clathrates are shown in Fig. 2 and Table 1 [21, 22]. They are Euler polyhedra, for which the equation $N_f + N_v = N_e + 2$ relates the numbers of faces (N_f), vertices (N_v), and edges (N_e).

The polyhedron that prevails in the clathrate structures is the 20-vertex pentagonal dodecahedron, $[5^{12}]$. It contains fivefold symmetry axes and cannot fill the space itself. Therefore, it combines with the polyhedra of other types in various clathrate structures. The popularity of the pentagonal dodecahedron in clathrates can be explained by the fact that the ideal bond valence angle in it (108°) is very close to the value of 109.45° expected for the tetrahedrally bonded network. As a consequence, most of the clathrate structures are built up by a combination of pentagonal dodecahedra with one or two other cage polyhedra. Evidently, the effective size of the polyhedral cage depends on the number of vertices and the nature of the element that forms the framework. Larger polyhedra require larger guests for achieving the stability of the clathrate structure. In most cases, a certain number of the polyhedra should be filled by the guest atoms. The atoms forming hexagonal faces have more distorted environment compared to the atoms belonging exclusively to pentagonal faces. Atomic positions in the vertices of the hexagonal faces are often occupied by nontetrel atoms, such as B or P, or vacant, thus releasing the stress.

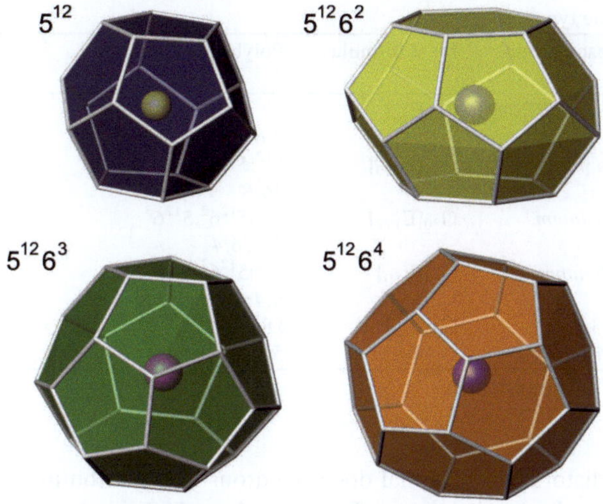

Fig. 2 Clathrate-forming polyherda

Table 1 Clathrate-forming polyherda

Polyhedron	Number of faces	Number of vertices	Observed in clathrate types
Pentagonal dodecahedron	5^{12}	20	I,II,III,IV,VIII[a], IX
Tetrakaidecahedron	$5^{12}6^{2}$	24	I, III, IV
Pentakaidecahedron	$5^{12}6^{3}$	26	III, IV
Hexakaidecahedron	$5^{12}6^{4}$	28	II

[a]Distorted 5^{12} polyhedron, see Sect. 3.5

2.2 *Zintl Clathrates Structure Types*

Zintl clathrates crystallize into six structure types. These types are denoted by Latin numbers; for instance, type I clathrate is the most common clathrate type, in which the [5^{12}] and [$5^{12}6^{2}$] polyhedra alternate in a 1:3 ratio. Historically, it was the first discovered type of the clathrate structures, and it is common for many clathrates-hydrates and Zintl clathrates. Three other types, type II, type III, and type IV, are also represented by both hydrates and Zintl phases, though for each of them the number of known compounds is much lower than for the type I (Table 2). There are no Zintl phases with the structures of types V, VI, and VII; these structures are found only for gas and liquid hydrates. Finally, there exist two other clathrate structural types denoted as clathrate-VIII and clathrate-IX that have no analogs among hydrates.

Table 2 lists all structure types pertaining to the Zintl clathrates, emphasizing the type and the number of polyhedra composing each structure. Noticeably, clathrate-VIII is the only clathrate type composed from just one kind of polyhedra, which is a

Table 2 Clathrate types

Clathrate type	Space group	Ideal formula	Polyhedra, *ratio*	Approx. number of phases
I	$Pm\bar{3}n$	$G_8[E_{46}]$	5^{12}:$5^{12}6^2$ *2:6*	150
II	$Fd\bar{3}m$	$G_{24}[E_{136}]$	5^{12}:$5^{12}6^4$ *16:8*	25
III	$P4_2/mnm$	$G_{30}[E_{172}]$	5^{12}:$5^{12}6^2$:$5^{12}6^3$ *10:16:4*	2
IV	$P6/mmm$	$G_{14}[E_{80}]$	5^{12}:$5^{12}6^2$:$5^{12}6^3$ *6:4:4*	1
VIII	$I\bar{4}3m$	$G_8[E_{46}]$	Distorted 5^{12}	4
IX	$P4_132$	$G_6[E_{25}]$	5^{12} + irregular polyhedra	20

G guest atom, *E* framework atom

considerably distorted pentagonal dodecahedron. The distribution of all clathrates over six structural types is uneven. There are about 150 compounds that crystallize into the dominating type I. More than 20 compounds represent each of the two other types, type II and type IX. The remaining types are very uncommon. The only example of the type IV clathrate is the intermetallic compound in the Li–Mg–Cu–Ga system [23]. The type III and type VIII clathrates are represented by two and four compounds, respectively.

According to the formal charge of the framework, all clathrates are classified as anionic, cationic, and neutral despite of their structure types. In the classical anionic clathrates, the framework bears a formal negative charge and the guest atoms are positively charged; they are the cations of alkali or alkaline-earth metals or europium. The reverse situation is observed in cationic clathrates (also known as inverse clathrates), where the framework has a formal positive charge and the guest atoms are the negatively charged halide or telluride anions. The unique example of the neutral (and guest-free) clathrate is the peculiar polymorph of germanium crystallizing with the type II clathrate structure [19].

3 Crystal Structure of Zintl Clathrates

3.1 Clathrate-I

Clathrate-I crystallizes in the cubic space group $Pm\bar{3}n$. The crystal structure of clathrate-I consists of two types of polyhedra: pentagonal dodecahedra [5^{12}] and tetrakaidecahedra [$5^{12}6^2$] (Fig. 3). Tetrakaidecahedra share their hexagonal faces forming a columns running along all three directions, [100], [010], and [001], thus building a 3D system. Smaller pentagonal dodecahedra are located in between such columns and become isolated from each other. The unit cell of the clathrate-I structure contains two small [5^{12}] and six large [$5^{12}6^2$] polyhedra. There are 46

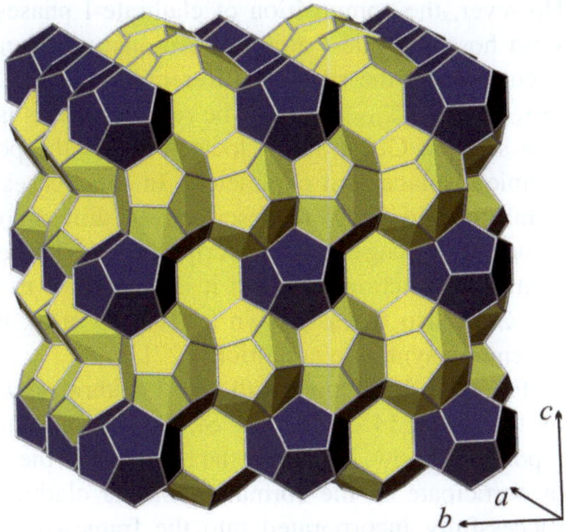

Fig. 3 Crystal structure of clathrate-I

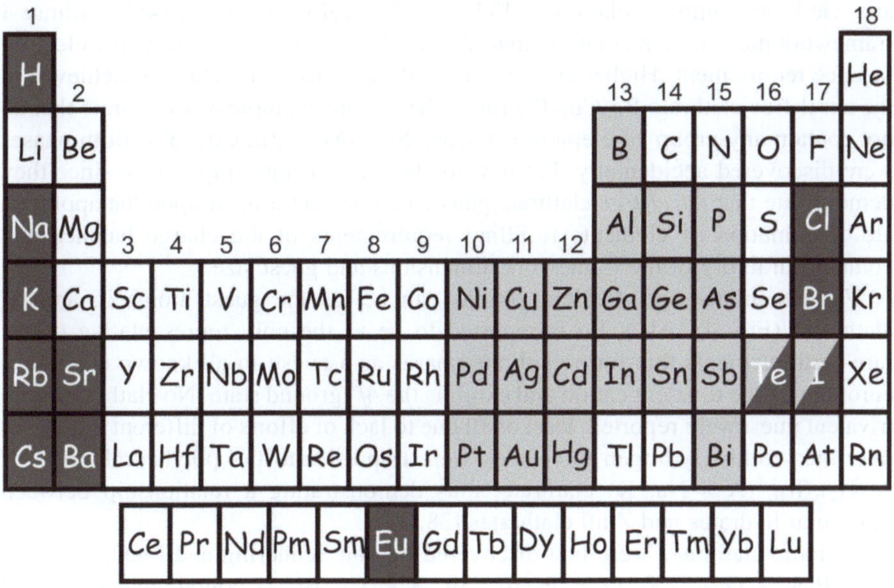

Fig. 4 Periodic system emphasizing Zintl clathrate forming elements: guest atoms – *dark gray*, host framework atoms – *light gray*

host framework atoms; therefore, its formula can be generalized as $G_2G'_6[E_{46}]$. Clathrate-I is the most abundant clathrate type and can be formed with a combination of different elements as shown in Fig. 4. Detailed compositions of known clathrates were reported elsewhere [22]. The simplest anionic clathrate is Na_8Si_{46}, where all the guest and host positions are 100% occupied by atoms of one sort [1].

However, the composition of clathrate-I phases can be much more complicated. Both host and guest atomic positions may contain vacancies and exhibit mixed occupation of the sites by different atoms, e.g., $Ba_{7.7}\square_{0.3}[Si_{42.7}Ge_{3.1}\square_{0.2}]$ [24] or $[Sn_{14}In_{10}P_{21.2}\square_{0.8}]I_8$ [25]. The reason for this is explained in the electronic structure section. Guest atoms located inside the polyhedral cages exhibit very large atomic displacement parameters. In some cases, the guest atoms inside the larger tetrakaidecahedra were described as shifted from their ideal position 6*d* in the center of the cage. Such description leads to better fit of experimental X-ray and neutron diffraction data, but it is applicable to only a limited number of systems [26–28]. Also it was shown by Iversen et al. that guest atom displacement may depend on synthetic conditions [27].

Frameworks of almost all type I clathrates (with only two exceptions) are based on the group 14 elements – Si, Ge, or Sn. No clathrates containing C or Pb are reported till now. However, the elements of the second and sixth periods are known to participate in the formation of the clathrate framework. Indeed, boron was successfully incorporated into the framework of $K_7B_7Si_{39}$ [29], while mercury forms $A_8Hg_xE_{46-x}$ (A = K, Rb, Cs; E = Ge, Sn) [30, 31]. Transition metals of the groups 10 and 11 participate in the formation of the clathrate frameworks of Si- and Ge-based anionic clathrates [32–35]. Typically, a tetrel-based clathrate-I framework may contain not more than six transition metal atoms due to the electron balance requirement. Higher concentration of the transition metal was achieved in the tetrel-free clathrate $Ba_8Cu_{16}P_{30}$ [36]. The second example of a clathrate that do not contain any group 14 element is $Cs_8M_{18}Sb_{28}$ (M = Zn, Cd) [37]. Both phases were discovered accidentally, but nevertheless are of high importance since they demonstrate that *tetrel-free* clathrate phases can be synthesized upon the appropriate combination of elements fulfilling requirements of the charge balance and commensurability of the framework dimensions and guest size.

Various mono- and divalent cations can act as the guest atoms in anionic clathrates (Fig. 4). Also, Eu is reported to be as the only representative of the rare-earth elements that form clathrate phases as a guest; in all known clathrates, europium is the divalent cation and exhibits the $4f^7$ ground state. No clathrates with trivalent guest were reported, least of all due to lack of efforts of different groups all over the world. Hydrogen molecules were trapped into the polyhedral cages of $A_x(H_2)_ySi_{46}$ (A = Na, K) clathrate, thus demonstrating a relationship between molecular hydrates and Zintl clathrates [38, 39].

Cationic clathrates were first discovered by von Schnering and Menke in 1972 [40]. Phases $Ge_{38}Z_8X_8$ (Z = P, As, Sb; X = Cl, Br, I) contain pnicogen and germanium atoms in the framework and halogen anions inside the polyhedral cages [41]. Later on, cationic clathrates containing tin and silicon were also produced [42–44]. Also, the guest chemistry was expanded and now includes Te [45–49]. Similar to anionic clathrates, the host framework of the cationic clathrates may contain vacancies as well as different elements of groups 11–17 [25, 50–54]. Tellurium and iodine are unique elements since they may serve as host and guest atoms simultaneously in the same clathrate structure; the examples are $[Si_{44.5}I_{1.5}]I_8$, $[Ge_{43.33}I_{2.67}]I_8$, and $[Si_{38}Te_8]Te_8$ [43, 45, 54].

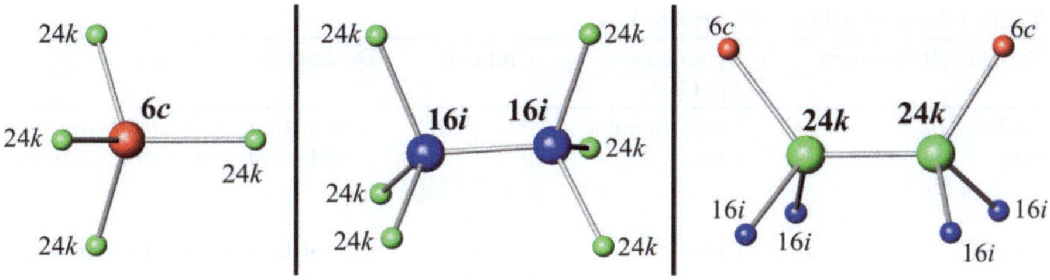

Fig. 5 Atomic environment of different framework positions in the clathrate-I structure

In the ideal clathrate-I structure, 46 framework atoms are distributed over three crystallographic positions: 6*c*, 16*i*, and 24*k*. Atoms in the two latter positions form both types of polyhedra, while the 6*c* position belongs only to the larger $[5^{12}6^{2}]$ polyhedra. The atom in the position 6*c* is tetrahedrally coordinated by four atoms occupying the 24*k* position; the atom in the 16*i* position is tetrahedrally coordinated by three atoms of the position 24*k* plus one atom of the position 16*i*; in turn, the atom in the position 24*k* is surrounded by one 6*c* atom, two 16*i* atoms, and one 24*k* atom (Fig. 5). Thus, the atoms in the positions 16*i* and 24*k* form *homoposition* bonds which determine the preferential occupation by heteroatoms or vacancies of the different crystallographic positions in the clathrate-I structure, which, in turn, may lead to the formation of the superstructures.

Several examples of the superstructures of clathrate-I were reported (Table 3). Note that the cases of the reduction of symmetry within the original clathrate cubic unit cell are also included in Table 3. Three ways of forming superstructures with the increasing of the unit cell volume are shown in Fig. 6.

In most cases, the superstructure formation is caused by a full or partial ordering of the vacancies in the positions of framework atoms. Another driving force is a segregation of atoms of quite different nature over different clathrate framework positions. For clathrate-I in the Si–P–Te system, the superstructure formation is caused by the partial ordering of guest tellurium vacancies inside smaller pentagonal dodecahedra, which is accompanied by the segregation of Si and P over different framework positions in such a way that phosphorus atoms form the cage around the vacancies in the guest Te positions [47, 48]. For $A_8Sn_{44}\square_2$ (A = K, Rb, Cs), the ordering of vacancies is a temperature-driven and reversible process: at room temperature, these phases crystallize in the ordered supercell of clathrate-I, while upon increasing the temperature the phase transition occurs leading to the disordering of vacancies and lowering symmetry to the primitive cubic clathrate-I unit cell [56–58, 60]. For $Ba_8Ge_{43}\square_3$, it was mentioned [55] that ordering of germanium vacancies also depends on the temperature treatment of the sample; however, no sharp phase transition was detected. Decomposition of $Ba_8Ge_{43}\square_3$ into Ba_6Ge_{25} and Ge was suggested to occur upon heating; however the temperature of such decomposition was controversial: 770 K [61] and 1,083 K [62]. The latter was confirmed by the recent detailed studies which demonstrated that $Ba_8Ge_{43}\square_3$ is a thermodynamically stable phase in the temperature range of 1,043–1,083 K [63].

Table 3 Superstructures of clathrate-I

Formula [Reference]	Symmetry Sp. Gr.	Unit cell	Description
$Ba_8[Cu_{16}P_{30}]$ [36]	Ortho-rhombic *Pbcn*	$a \approx \sqrt{2}a_o$ $b \approx a_o$ $c \approx 2\sqrt{2}a_o$ $V \approx 4V_o$	Segregation of Cu and P over different framework positions
$Ba_8[Ge_{43}\square_3]$ [55]	Cubic $Ia\overline{3}d$	$a \approx 2a_o$ $V = 8V_o$	Complete ordering of vacancies within the framework
$A_8[Sn_{44}\square_2]$ A = Rb, Cs [56–58]	Cubic $Ia\overline{3}d$	$a \approx 2a_o$ $V \approx 8V_o$	Partial ordering of vacancies within the framework
$[Ge_{38}Z_8]X_8$ Z = P, As, Sb; X = Cl, Br, I [40, 41]	Cubic $P\overline{4}3n$	Similar to original	Segregation of Ge and Z over different framework positions
$[Sn_{14}In_{10}P_{21.2}\square_{0.8}]I_8$ [25]	Tetragonal $P4_2/m$	$a \approx \sqrt{5}a_o$ $c \approx a_o$ $V \approx 5V_o$	Partial ordering of P vacancies within the framework
$[Sn_{20.5}\square_{3.5}As_{22}]I_8$ [59]	Cubic $Fm\overline{3}$ or $F23$	$a \approx 2a_o$ $V \approx 8V_o$	Partial ordering of Sn vacancies within the framework
$[Si_{38}Te_8]Te_8$[a] [45, 46]	Cubic $P\overline{4}3n$	Similar to original	Segregation of Si and Te over different framework positions
$[Si_{46-x}P_x]Te_{8-y}\square_y$ [47, 48]	Cubic $Pm\overline{3}$	Similar to original	Partial ordering of Te vacancies, segregation of Si and P over framework positions

□ denotes vacancy
[a] $R3c$ distortion of the same clathrate type was also reported [45, 46]

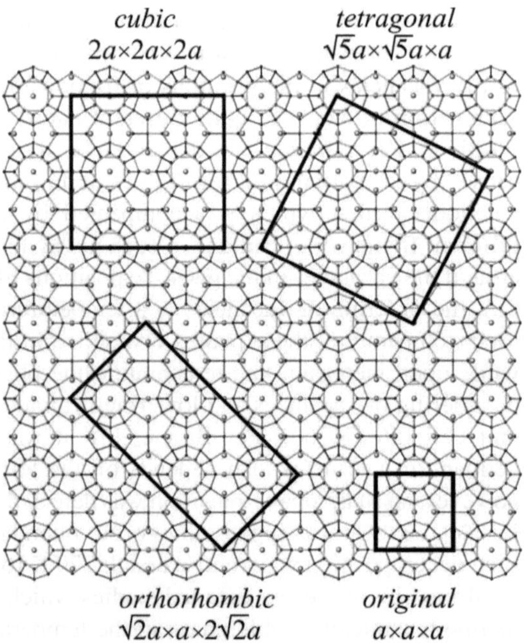

Fig. 6 The choice of the unit cell for three variants of superstructure of clathrate-I

Upon cooling $Ba_8Ge_{43}\square_3$ forms at 1,083 K as the ordered phase by the peritectic reaction from α-Ge and the melt and crystallizes at 1,081 K in the eutectic mixture with Ba_6Ge_{25}.

3.2 *Clathrate-II*

Pentagonal dodecahedra [5^{12}] together with hexakaidecahedra [$5^{12}6^4$] form the crystal structure of clathrate-II (cubic space group $Fd\bar{3}m$). Larger [$5^{12}6^4$] polyhedra share all their hexagonal faces forming a diamond-like 3D network. Each hexakaidecahedra is surrounded by 16 pentagonal dodecahedra, forming a supertetrahedron (Fig. 7). In this way, a complete filling of the space is achieved. The unit cell of clathrate-II contains 16 small and 8 large polyhedra. The formula for 136 host framework atoms and 24 guests of two different types can be generalized as $G_{16}G'_8[E_{136}]$. Only anionic clathrates-II were reported so far, and most of them are compounds with the frameworks made exclusively of Si or Ge atoms, in which the guests are alkali metals, e.g., $Na_x[Si_{136}]$ or $Na_{16}Rb_8[Ge_{136}]$ [1, 21]. Two known examples of framework-substituted clathrates-II include gallium and transition metal compounds: $Ba_{16}\square_8[Ga_{32}Sn_{104}]$ [64] and $Na_{16}Cs_8[M_xGe_{136-x}]$ M = Cu, Ag [65, 66]. Often, only one type of cages is preferably/exclusively filled in clathrate-II due to the large difference in the size of the clathrate-forming polyhedra. For instance, potassium atoms in $K_{8.6}Ge_{136}$ preferentially occupy larger 28-vertex polyhedra [67]. On the other hand, in the larger tin clathrate-II framework of

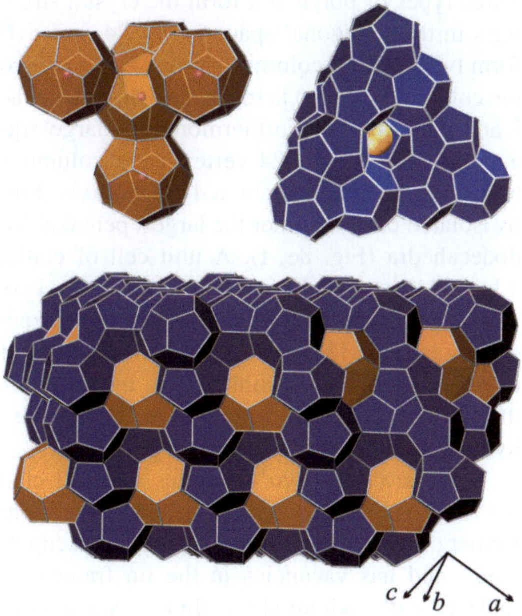

Fig. 7 *Top left*: $(5^{12}6^4)_5$ tetrahedron; *top right*: $(5^{12})_{16}$ supertetrahedron; *bottom*: general view of the crystal structure of clathrate-II

$Ba_{16}\square_8[Ga_{32}Sn_{104}]$, barium atoms exclusively fill small pentagonal cages while larger $[5^{12}6^4]$ polyhedra remain vacant [64]. Filling of both types of polyhedra can be achieved in clathrates-II with two different types of guests. For instance, the combination of Cs and Na produces a completely filled clathrate-II structure of $Na_{16}Cs_8[E_{136}]$ (E = Si, Ge), where Cs resides inside larger hexakaidecahedra and Na is trapped by pentagonal dodecahedra [68]. Interestingly, the clathrate framework contracts if sodium atoms fill exclusively the 28-vertex cages of Na_xSi_{136}; however, when Na starts to enter the smaller 20-atom cages, the lattice begins to expand [69].

The unique feature of the clathrate-II compounds is the ability to persist the extraction of the guest atoms. New allotropic modification of germanium – clathrate-II Ge_{136} – was prepared by a mild oxidation of Na_4Ge_9 in ionic liquids [19]. By the combination of X-ray diffraction, electron microscopy, and EEL spectroscopy, as well as physical properties measurements, it was shown that the discussed clathrate-II is guest-free. Silicon-based clathrate-II Na_xSi_{136} can be prepared by thermal decomposition of NaSi binary phase [1]. Annealing of Na_xSi_{136} with iodine allowed to drastically reduce sodium content, x. Based on ICP and EPR results, the authors claimed to produce an Na_xSi_{136} phase with $x = 0.0058$ [18].

Chemistry and physics of anionic clathrate-II compounds were recently reviewed by Beekman and Nolas [70].

3.3 Clathrate-III

Three types of polyhedra form the crystal structure of clathrate-III, which crystallizes in the tetragonal space group *P*4$_2$*/mnm* (Fig. 8g). Tetrakaidecahedra $[5^{12}6^2]$ form two types of columns along [001], single or four-condensed. The first type of the columns (Fig. 8a) is formed by sharing the hexagonal faces, similar to clathrate-I and clathrate-IV. Furthermore, the large square-shaped channels are filled by another type of the 24-vertex cage column (Fig. 8d) formed by a couple of tetrakaidecahedra lying on a 4_2 screw axis (Fig. 8c). The remaining space is filled by isolated pairs of either the largest pentakaidecahedra $[5^{12}6^3]$ or the smallest $[5^{12}]$ dodecahedra (Fig. 8e, f). A unit cell of clathrate-III contains 10 smallest $[5^{12}]$, 4 largest $[5^{12}6^3]$ and 16 medium-size $[5^{12}6^2]$ polyhedra. There are 172 host framework atoms, and the formula can be generalized as $G_{10}G'_{16}G''_4[E_{172}]$. Note that the chiral clathrate-IX is sometimes incorrectly referred to as clathrate-III, probably due to rarity of Zintl clathrates-III. In fact, in the hydrates classification by Jeffrey the term clathrate-III was used to define the crystal structure type of bromine hydrate more than two decades ago [6].

There are only two examples of Zintl clathrate-III: one anionic $A_{30}[Sn_{172-x}Na_y\square_z]$ (A = Cs, Cs/Rb) [71] and one cationic $[Si_{172-x}P_x]Te_{2x}\square_{30-2x}$ [72]. The former clathrate-III exhibits a complete occupation of all polyhedral cages by alkali atoms and has vacancies in the tin framework. Part of the missing tin pairs is replaced with sodium atoms. In opposite, cationic clathrate-III exhibits a complete

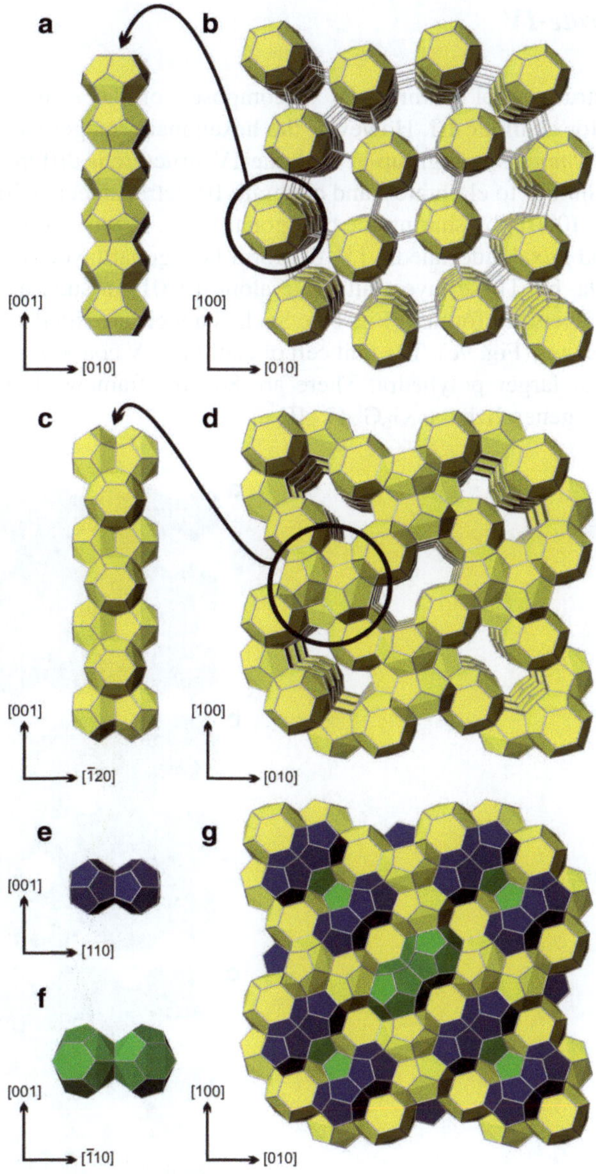

Fig. 8 Crystal structure of clathrate-III: arrangements of $5^{12}6^2$ (**a–d**), 5^{12} (**e**), $5^{12}6^3$ (**f**), and general view of the structure (**g**)

occupation of all framework positions by Si and P atoms together with the presence of vacancies in the guest Te sites: small pentagonal cages are partially occupied by Te, similar to Si–P–Te clathrate-I [47, 48]. Owing to varying occupancy of Te atoms inside the 5^{12} polyhedra, both Si–P–Te clathrates, of types I and III, exhibit a relatively wide homogeneity range which is not typical for Zintl phases.

3.4 Clathrate-IV

The crystal structure of clathrate-IV is composed of the same three types of polyhedra as for clathrate-III. However, the hexagonal arrangement of such polyhedra in the *P*6/*mmm* space group in clathrate-IV structure is different from that in clathrate-III. Similar to clathrate-I and clathrate-III, tetrakaidecahedra [$5^{12}6^2$] form columns along [0001] by sharing the hexagonal faces. In turn, pentagonal dodecahedra [5^{12}] and hexakaidecahedra [$5^{12}6^3$] form hexagonal layers perpendicular to [0001] (Fig. 9a, b). These layers alternate along [0001] in such a way that large hexagonal channels are formed (Fig. 9d). Such channels are filled with columns of the 24-vertex cages (Fig. 9c). The unit cell of clathrate-IV contains 6 smallest and 4 of each type of larger polyhedra. There are 80 host framework atoms, and the formula can be generalized as $G_6G'_4G''_4[E_{80}]$.

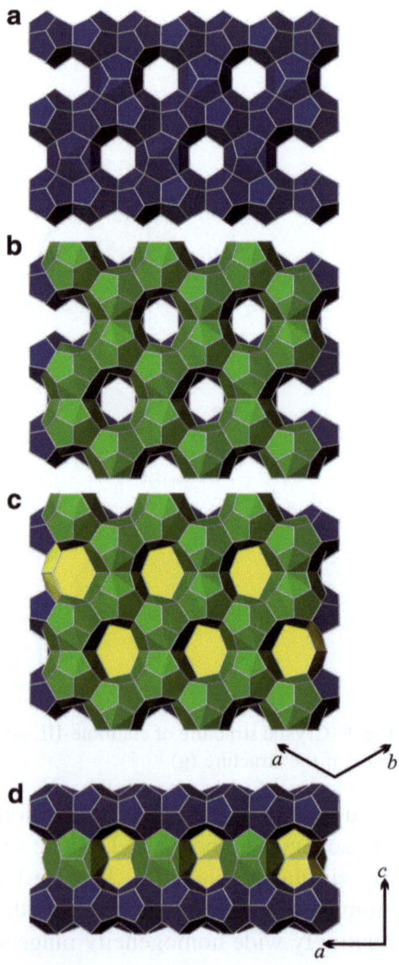

Fig. 9 Crystal structure of clathrate-IV: hexagonal layers of 5^{12} (**a**) and $5^{12}6^3$ (**b**) polyhedra, filling channels with $5^{12}6^2$ columns (**c**), general view (**d**)

An interesting feature of clathrate-IV is that no compounds crystallizing in the ideal space group *P6/mmm* were reported. Lowering of the symmetry preserving the general polyhedral arrangements is typical for type IV hydrates [6]. Only one known representative of intermetallic clathrates-IV – $Li_{14.7}Mg_{36.8}Cu_{21.5}Ga_{66}$ – follows this tendency and crystallizes in the $P\bar{6}m2$ space group [23]. This phase features different Ga, Li/Cu/Ga, and Li/Mg/Cu/Ga clusters situated inside the clathrate polyhedral cages formed by Li and Mg atoms. This distinguishes $Li_{14.7}Mg_{36.8}Cu_{21.5}Ga_{66}$ from all others clathrates discussed in this chapter. In all other Zintl clathrates, the polyhedra edges correspond to covalent chemical bonds between framework atoms, with exceptions of a vacancy in the polyhedral vertex. Contrasting, the distances in the Li/Mg clathrate framework are longer than 2.9 Å and, thus, cannot be described as covalent bonds.

3.5 Clathrate-VIII

Clathrate-VIII is a polymorph modification of clathrate-I, i.e., each clathrate-VIII (α-phase) upon heating transforms into clathrate-I (β-phase) of the same composition. The crystal structure of clathrate-VIII consists of only one type of polyhedral cages (Fig. 10, top), which can be described as strongly distorted pentagonal

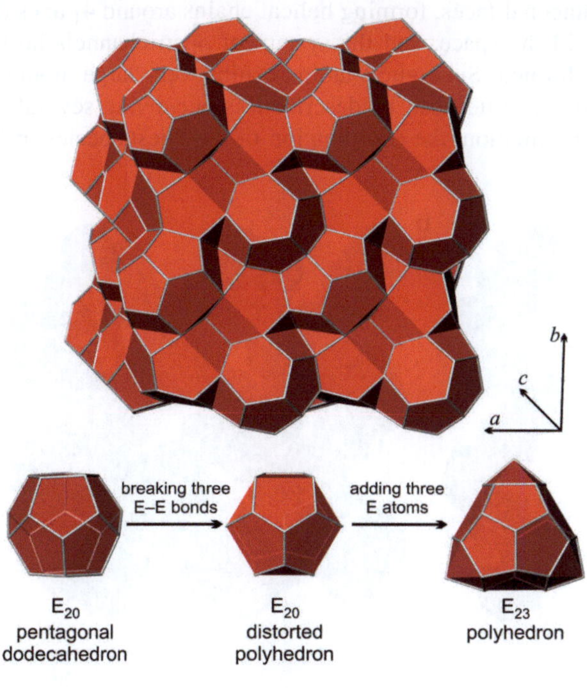

Fig. 10 Crystal structure of clathrate-VIII (*top*), transformation of pentagonal dodecahedron into E_{20+3} polyhedron (*bottom*)

polyhedra, E_{20+3}. Such a polyhedron can be derived from the regular pentagonal dodecahedron by breaking three edges and adding three new vertices, thus forming nine new edges (Fig. 10, bottom). Filling of all the space with such E_{23} polyhedra is impossible, and small vacant eight-vertex polyhedral cages are also present in the crystal structure. Clathrate-VIII crystallizes in the cubic space group $I\bar{4}3m$, with the general formula $G_8[E_{46}]$. There are only four examples of Zintl clathrates-VIII documented in the literature, all of them are anionic clathrates: $Eu_8[Ga_{16}Ge_{30}]$, $Ba_8[Ga_{16}Sn_{30}]$, $Sr_8[Al_xGa_{16-x}Si_{30}]$, and $Sr_8[Al_xGa_{16-x}Ge_{30}]$ [73–76].

Using the Eu-containing compound as an example, the structural relationship between the high-temperature clathrate-I phase and the low-temperature clathrate-VIII phase was investigated with a topological approach based on periodic minimal surfaces approximants. The description of the structural transition was formulated in terms of a global transformation of the two four-connected covalent frameworks, and appears as a rotation around a distinct cubic threefold axis [77].

3.6 Clathrate-IX or Chiral Clathrate

The crystal structure of clathrate-IX (space group $P4_132$) is built from only one type of regular polyhedra, the pentagonal dodecahedron [5^{12}] (Fig. 11). These polyhedra share their pentagonal faces, forming helical chains around 4_1 axes (Fig. 11a). The chains do not fill the space, and the system of open channels hosting additional guest atoms is formed. Such guest atoms exhibit very large atomic displacement parameters and very often can be described as occupying several split positions. In the first approximation, the coordination of the guest species inside these open

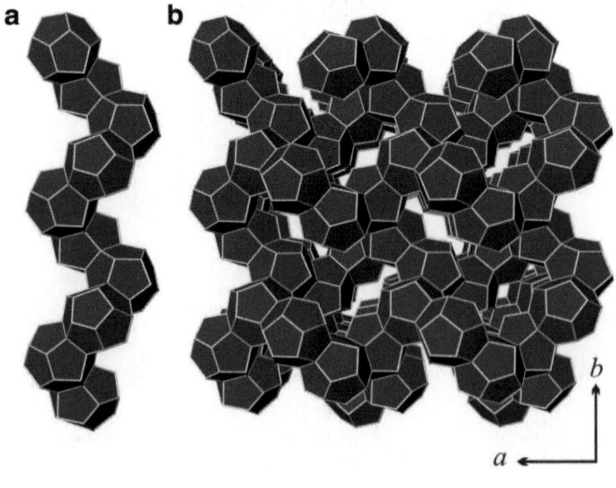

Fig. 11 Crystal structure of clathrate-IX: helical chain of pentagonal dodecahedra (**a**), general view along [001] (**b**)

channels can be described as follows: one atom is situated inside the open 20-vertex polyhedron, while the second one is trapped inside the strongly distorted cube. Generally, one can conclude that the clathrate-IX structure type exhibits some flexibility in the amount of guest atoms and in the number of their positions. For instance, silicon clathrate-IX $Ba_6[Si_{25}]$ has only six guest atoms [78]. The same is true for Ge-containing clathrates-IX: $Ba_6[Ge_{25}]$ [79, 80], $Na_2Ba_4[Ge_{25}]$ [81], and $Ba_6[Ge_{25-x}M_x]$ (M = In, Sn, $x \leq 4$) [82, 83]. In turn, the tin-based framework has larger channels and is able to accommodate more guests, up to 7.4 per formula as in $K_{7.4}[Sn_{25}]$ [84]. It should be noted that clathrate-IX $K_6[Sn_{25}]$ was also reported [85]. A combination of two different types of guest atoms allows to maximize the number of guests in clathrate-IX; an example is provided by $Rb_5Na_3[Sn_{25}]$ [86]. Heteroatomic substitution in the tin framework results in the formation of the Zintl clathrate with the heaviest element, bismuth: $K_6[Sn_{23}Bi_2]$ [85]. No cationic type IX clathrates were reported so far. Another important feature of clathrate-IX is that the framework contains three-connected atoms; this is not observed for other clathrates.

The comprehensive study of $Ba_6[Ge_{25}]$ by Grin et al. reveals that the displacement of the guest atoms inside the open channels may affect the structure of the framework. Variable temperature diffraction experiments indicate that the displacement of one of the Ba atoms occurs without changing the space group and is accompanied by breaking one Ge–Ge bond in the framework. This observation correlates well with the observed changes in physical properties [87].

Another detailed study of this structure type was performed by von Schnering et al. by means of periodic nodal surfaces. They show that the clathrate-IX structure can be divided into two nonintersecting partial substructures, one of them containing pentagonal polyhedra and the other consisting of open channels [88]. In the same work, a summary of known clathrates-IX and their guest atomic displacement is given.

4 Clathrates as Zintl Phases

4.1 Application of the Zintl Electron-Count Scheme

The composition and transport properties of the vast majority of Zintl clathrates can be explained and, most significantly, predicted based on the Zintl electron-count scheme. Two basic ideas of the Zintl concept [89, 90] are applicable to clathrates, which are (1) the complete charge transfer between the framework and guest atoms, and (2) the realization of the electron octet by each atom. For clathrates with only four-bonded framework atoms (all types except clathrate-IX), it means that four electrons are required per each framework atom. Thus, for each clathrate structure type one can calculate the electron-balanced composition. A guest-free tetrel framework $[E_z]$ is charge-balanced and has $4z$ framework electrons. The presence of the guest would result in either donation (guest cations in conventional anionic

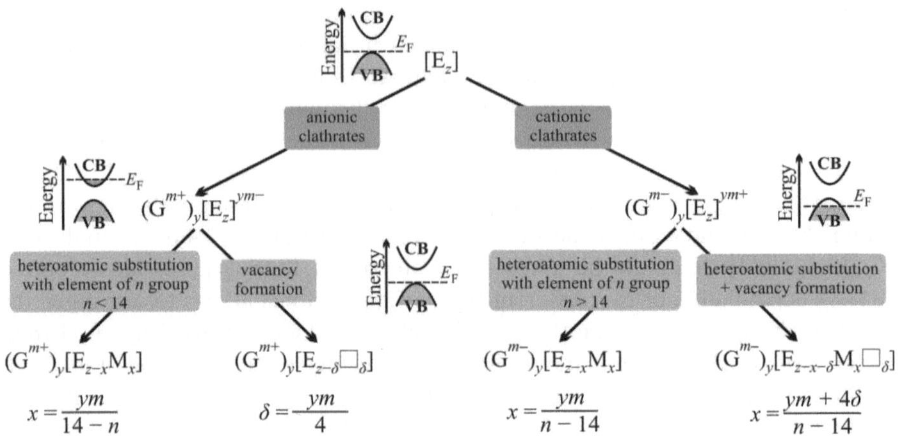

Scheme 1 Application of Zintl rules to clathrates structures. *CB* conduction band, *VB* valence band

clathrates) or withdrawal (guest anions in inverse or cationic clathrates) of electrons to/from the framework. The required number of the framework electrons, 4*z*, can be maintained either by substitution of tetrel elements with a heteroatom having a different number of valence electrons or by formation of vacancies in the framework (Scheme 1). According to the Zintl concept, the formation of the vacancies in the clathrate framework results in generation of three-coordinated (3*b*) atoms with localized electron pairs. Each 3*b* atom requires one additional electron compared to the "normal" four-bonded atom. One vacancy creates four three-coordinated atoms; consequently, formation of one vacancy would compensate four excessive electrons. Localized electron pairs correspond to nonbonding E–E states and significantly contribute to the states at the top of the valence band; therefore, vacancy formation determines the band gap and transport properties of both anionic and cationic clathrates [25, 90, 91]. Formation of vacancies alone is not able to produce electron-balanced cationic clathrate compositions and is always combined with heteroatomic substitutions [22]. Scheme 1 describes only ternary clathrate compositions; however, it can be extended to quaternary and higher cases, for example, by introducing two heterometals of different groups from the Periodic Table and/or a mixture of different guest atoms.

In terms of the band structure, the Zintl approach implies that the bands arising from guest cations (in classical anionic clathrates) are located well above the framework bands, while the bands coming from guest anions (in inverse or cationic clathrates) lie lower in energy compared to framework bands. Moreover, as a rule, the guest and host bands do not mix significantly (Scheme 2). This is, without doubts, an idealized description; in real compounds some mixing of guest and host bands occurs. For example, in $Ba_{8-\Delta}[Si_{46}]$ mixing of Ba and Si orbitals is relatively strong and results in the superconducting properties of such compounds [13, 14, 92].

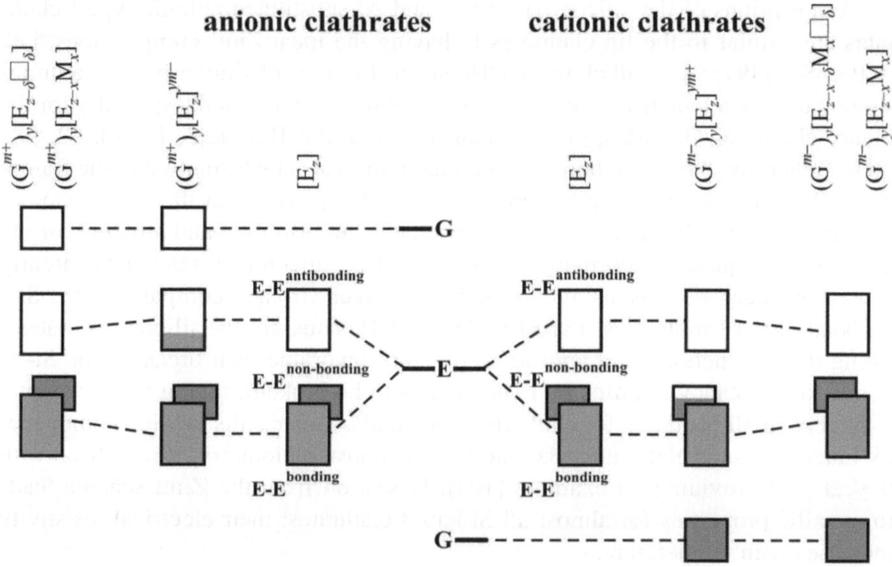

Scheme 2 Band structure of Zintl clathrates

Clathrates with the composition complying with the Zintl scheme are expected to exhibit semiconducting properties and a narrow homogeneity range. Deviations from the Zintl scheme typically result in metallic-like behavior of electrical resistivity; besides, such phases often exhibit a wide homogeneity range. Examples of clathrate compositions, limits of applicability, and resulting physical properties are considered below separately for different clathrate types.

4.2 *Clathrate-I*

4.2.1 Anionic Clathrate-I

Clathrate-I has 46 framework atoms; thus, $46 \times 4 = 184$ electrons per formula unit are necessary to maintain the charge balance. Tin-based clathrates-I are Zintl phases in most cases. The Sn_{46} framework would be charge balanced. Hypothetical clathrates $A_8[Sn_{46}]$ (A = alkali metal, $m = 1, y = 8$) have $ym = 8$ extra electrons from guest cations. These can be compensated either by heteroatomic substitution with, for example, Ga ($n = 13$, $x = 8$) or Zn ($n = 12$, $x = 4$) resulting in $A_8[Ga_8Sn_{38}]$ [93, 94] and $A_8[Zn_4Sn_{42}]$ [95], or by creation vacancies in the framework, $\delta = 2$, leading to the formation of the $A_8[Sn_{44}\square_2]$ phases [56–58, 60]. Divalent guests, such as barium, require more heteroatoms in the framework, e.g., $Ba_8[Ga_{16}Sn_{30}]$: $m = 2, y = 8, ym = 16$ [96].

Compositions of the silicon-based Ga- and Al-substituted anionic type I clathrates are similar to the tin clathrates in having the ideal Zintl compositions, i.e., $A_8[Ga_8Si_{38}]$ [97] (A = alkali metal). However, for most of silicon-based clathrates significant deviations from the Zintl count scheme were reported, especially for the binary clathrates $G_{8-\Delta}[Si_{46}]$ (G = alkali metal and/or Ba) [1, 2, 13, 14, 92, 98–103]. Generally, the Si-containing clathrates avoid vacancy formation in the framework. This demonstrates limitations of the Zintl approach and the importance of considering not only the formal oxidation states but also the band structure of the compound in question. Formation of vacancy-free silicon clathrates is apparently caused by energetic reasons: the Si–Si bond is much stronger compared to the Sn–Sn bond (226 kJ mole^{-1} vs. 151 kJ mole^{-1}) [104]. Thus, for the silicon clathrates-I filling the conduction band appears to be more favorable than breaking the Si–Si bonds (one vacancy requires four such bonds to be broken), taking into account a relatively small band gap for clathrate compounds. This results in wide homogeneity ranges for such clathrates, viz. guest atoms nonstoichiometry, $Rb_{8-\Delta}[Si_{46}]$ with $0 \leq \Delta \leq 2$ providing an example [103]. Deviation from the Zintl scheme leads to metallic properties for almost all Si-based clathrates; their electrical resistivity increases with temperature.

As was mentioned above, $Ba_{8-\Delta}[Si_{46}]$ displays superconducting properties below 8 K [13, 14, 92, 101, 102]. The superconducting transition temperature T_C strongly depends on the Ba content, i.e., the highest T_C was obtained for $Ba_8[Si_{46}]$. The authors propose that the superconducting state is realized due to the mixing of Ba and Si orbitals. The increased amount of Ba results in higher degree of such mixing [92].

Vacancy formation in the guest sublattice does not necessarily result in the deviation from the Zintl count. In the clathrate-I compound $K_7\square_1[B_7Si_{39}]$, seven extra electrons from potassium are compensated by the substitution of seven silicon atoms with boron. This substitution leads to the chemical compression of the framework; now it can accommodate only seven large potassium guests. Investigation of the transport properties confirms that $K_7\square_1[B_7Si_{39}]$ is a diamagnet and a semiconductor, which is typical for Zintl phases [29].

Germanium-based clathrates-I, unlike silicon clathrates-I, often exhibit a vacancy-containing framework, though the composition of some Ge-containing clathrates deviates from the Zintl electron-count scheme [22]. Similar to tin clathrates, there is a group of substituted germanium clathrates with a general formula $A_8[M_8Ge_{38}]$ (A = K, Rb, Cs; M = Al, Ga, In) [93, 94, 105–107]. More complex compositions, featuring the presence of both a heterometal and a vacancy, were also reported for germanium, for instance, $Ba_8[Cd_4Ge_{40}\square_2]$ [108], where according to Scheme 1: $ym = 4(14–n) + 4\delta$; $ym = 16$, $n = 12$, and $\delta = 2$. Tiny deviations of the composition of the Ge-based type I clathrates from the Zintl scheme may result in either changing the carrier type from electrons to holes or switching the conductivity type from semiconductor-like to metallic. This is especially true for $A_8[Ga_{16}Ge_{30}]$ (A = Sr, Ba, Eu) where even the synthetic pathway to the target phase may significantly affect the Ga/Ge ratio and, correspondingly, transport properties [26–28]. Formation of the vacancies in the germanium framework is energetically more favorable compared to the silicon case (Ge–Ge bond, 188 kJ mole^{-1}) [104]; however, binary Ba–Ge clathrate-I exhibits significant deviation from the Zintl scheme. Indeed, the Zintl composition for this clathrate ($ym = 16$, $\delta = 4$) is $Ba_8[Ge_{42}\square_4]$, but such a phase does not exists, instead $Ba_8[Ge_{43}\square_3]$ is formed [55, 61, 63, 109]. As expected, $Ba_8[Ge_{43}\square_3]$ exhibits the metallic-like temperature dependence of resistivity in the range of 2–300 K [63]. Quantum-chemical calculations reveal the density-of-state (DOS) similar to those shown in Scheme 2: the DOS exhibits the pronounced band gap of ~0.2 eV, and the Fermi level is shifted to the conduction band due to the presence of "excessive" electrons [63]. Full ordering of the vacancies in $Ba_8[Ge_{43}\square_3]$ (vide supra) allows performing the state-of-the-art calculations for the visualization of chemical bonding, using the electron localizability indicator (ELI). It shows the presence of the covalent bonds between the four-bonded germanium atoms and the localization of electron pairs on the three-bonded germanium atoms surrounding the vacancy [63], thus fully confirming Zintl ideas concerning bonding in polar intermetallics.

In the anionic clathrates, vacancies are always observed in the 6*c* positions that form hexagonal rings of tetrakaidecahedra. When the vacancy is formed, the surrounding atoms in the 24*k* position relax, alternating the bond lengths and angles (Fig. 12b). Phases with a high concentration of vacancies, such as $A_8[Sn_{44}\square_2]$ ($A = Rb$, Cs) and $Ba_8[Ge_{43}\square_3]$, exhibit partial or full ordering of the vacancies along the 4_1 screw axis in the space group $Ia\overline{3}d$ with the $2 \times 2 \times 2$ supercell. In the case of $Ba_8[Ge_{43}\square_3]$, three vacancies per formula unit imply that each hexagonal ring contains one vacancy. The ideal Zintl phase "$Ba_8[Ge_{42}\square_4]$" requires 1/3 of the hexagonal rings to have two vacancies, which would result in the unreasonably long Ge–Ge distance between atoms located in the 24*k* position, leading to the two-bonded atoms and thus to the destabilization of the framework (Fig. 12c). This may be a reason for the deviation from the Zintl scheme in the case of $Ba_8[Ge_{43}\square_3]$. Thus, though the bonding in the clathrates can be well described exploiting the Zintl approach, the peculiarities of the crystal and band structures should be taken into account in each particular case.

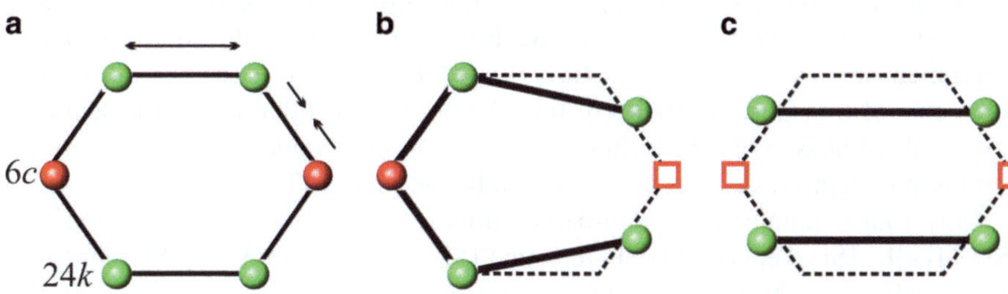

Fig. 12 Structure alternation upon vacancy formation in the 6*c* position of clathrate-I: (**a**) vacancy-free hexagonal ring; (**b**) formation of one vacancy; (**c**) formation of two vacancies. Note that formation of two vacancies results in unreasonably long distances between the neighboring 24*k* atoms

In general, heteroatoms of the groups 10–13 of the periodic system preferentially occupy the 6*c* position of the anionic clathrate-I framework [22]. For a long time, gallium was considered as an exception to this rule. The Ga/Ge distribution can hardly be elaborated by means of conventional X-ray diffraction techniques since these elements are the neighbors in the periodic system. Moreover, Ga and Ge have similar neutron scattering lengths, 7.3 fm and 8.1 fm, respectively. Early single crystal neutron diffraction experiments did not reveal any preferences in the occupation of different framework sites by gallium [110–112]. However, a combination of the state-of-the-art neutron diffraction with resonant synchrotron X-ray diffraction allowed distinguishing between Ga and Ge [26]. It was shown that the Ga atoms have a clear preference for the 6*c* site but avoid the 16*i* site. In contrast, antimony, the group 15 element, preferably occupies the 24*k* position in $Ba_8Ga_{18}Sb_2Ge_{26}$ [113, 114]. In the clathrate-I crystal structure, the position 6*c* exhibits the maximal deviation from tetrahedral symmetry, thus becoming the most "frustrated" framework position. Location of vacancies or heteroatoms in this position would release the stress, which explains the preferential occupancy of 6*c* position in the substituted and/or vacancy-containing clathrates-I.

4.2.2 Cationic Clathrate-I

Cationic clathrates-I were discovered by von Schnering and Menke almost four decades ago [40]. Halogens are the guest atoms in $[Ge_{38}Z_8]X_8$ (Z = P, As, Sb; X = Cl, Br, I) clathrates, and for achieving the electron balance the part of germanium atoms is substituted with pnicogen atoms (Scheme 1): $ym = 8, n = 15, x = 8$ [40, 41]. Germanium cationic clathrates-I containing tellurium ($n = 16$) and iodine ($n = 17$) in the framework were subsequently reported [53, 54]. The latter one has the Zintl composition, $[Ge_{43.33}I_{2.67}]I_8$: $ym = 8$, $x = 8/3$ with iodine atoms in two different oxidation states, I^- in the guest positions and I^{3+} in the host framework. The presence of two different types of iodine atoms was confirmed by means of X-ray photoelectron spectroscopy (XPS) [54]. The Te-containing clathrate has more complicated composition featuring the presence of both tellurium and vacancies in the framework, $[Ge_{40}Te_{5.3}\square_{0.7}]I_8$. This composition corresponds to the balanced Zintl composition $[Ge_{40}Te_{5.33}\square_{0.67}]I_8$ within the accuracy of the site occupancies determination: $x(n - 14) = ym + 4\delta$; $ym = 8$, $\delta = 2/3$, $n = 16$, $x = 32/6$. Measurements of transport properties of this phase confirmed that this is the semiconducting Zintl phase [53]. Tellurium is also able to play a guest role in the cationic germanium clathrates as was demonstrated by $Ge_{30+x}P_{16-x}Te_{8-\Delta}$ [49].

Only four examples of silicon-based cationic clathrates-I are known, they are $[Si_{44.5}I_{1.5}]I_8$, $[Si_{38}Te_8]Te_8$, $[Si_{40}P_6]I_{6.5}$, and $[Si_{46-x}P_x]Te_{8-\Delta}$ [43–48]. Similar to the anionic silicon clathrates, there are no vacancies in the silicon framework. Instead, the two latter phases exhibit vacancies in the guest positions located inside the smallest $[5^{12}]$ dodecahedra. The composition of only one clathrate, $[Si_{38}Te_8]Te_8$, complies with the Zintl scheme: $ym = 16$, $n = 16$, $x = 8$; however, for this phase studies of homogeneity range and transport properties were not performed [45, 46].

$[Si_{40}P_6]I_{6.5}$ has the composition lying very close to the Zintl one, "$[Si_{39.5}P_{6.5}]I_{6.5}$." Taking into account the low accuracy of the X-ray determination of the Si/P ratio in the same atomic positions, this compound may well obey the Zintl rule. The study of the homogeneity range indicates that $[Si_{40}P_6]I_{6.5}$ is a line phase; however, its transport properties were not investigated [44]. $[Si_{46-x}P_x]Te_{8-\Delta}$ exhibits a wide homogeneity range because the alteration of the tellurium content causes the change of the Si/P ratio. For this phase, $ym = 2(8-\Delta)$ and $(n-14) = 1$; thus, x should equal to $2(8-\Delta)$. The detailed studies of $[Si_{46-x}P_x]Te_{8-\Delta}$ show the slight compositional deviation from the Zintl concept, $x \leq 2(8-\Delta)$, which leads to the properties alien to the Zintl phases. Indeed, the samples with different x and Δ are the heavily doped semiconductors and temperature-dependent diamagnets [48].

The composition of all tin-based cationic clathrates-I is in accordance with the Zintl count, which is similar to the situation with their anionic counterparts. In contrast to the Ge-containing analogs, tin-based clathrates contain "excess" of pnicogen atoms in their frameworks that is compensated by either vacancies or heterometals with $n < 14$. For the P-containing frameworks, the vacancy formation leads to ternary phases $[Sn_{24}P_{19.2}\square_{2.8}]X_8$ (X = Br, I) with the Zintl composition [42, 91, 115]. According to Scheme 1, $ym = 8$, $\delta = 2.8$, $n = 15$; $x = (8 + 4 \times 2.8)/(15-14) = 19.2$. In the crystal structure of this compound, tin is located in the 24k position, while phosphorus occupies both 16i and 6c positions. Vacancies are formed in the phosphorus sublattice, exclusively in the 6c positions. The presence of vacancies leads to the splitting of the 24k position into two with different local environment (Fig. 13a, b). Ternary As-containing clathrate-I also

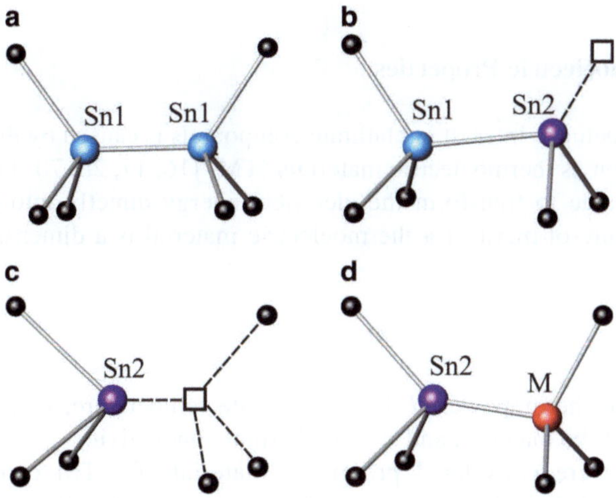

Fig. 13 Splitting of the tin atomic positions in the cationic clathrates. 24k positions are split into two (ternary phases) or three (quaternary phases) positions: Sn1, Sn2, and (optionally) M. In all cases, the Sn1 atoms are bound to themselves (**a**), in $[Sn_{24}P_{19.2}\square_{2.8}]I_8$ Sn2 is bound to Sn1 and phosphorus vacancy (**b**), in $[Sn_{20.5}\square_{3.5}As_{22}]I_8$ Sn2 is coordinated by only three As atoms (**c**), while in $[Sn_{17}Zn_7P_{22}]I_8$ and $[Sn_{19.3}Cu_{4.7}As_{22}]I_8$ Sn2 is bound to the heterometal, M = Zn or Cu (**d**). Pnicogen atoms are shown by *black spheres*, □ denotes vacancy

exhibits vacancies but they are found in the tin sublattice, in the 24*k* position [59]. The composition of $[Sn_{20.5}\square_{3.5}As_{22}]I_8$ corresponds to the Zintl scheme: $ym = 8$, $\delta = 3.5$, $n = 15$; $x = (8 + 4 \times 3.5)/(15–14) = 22$. The As-containing phase also features the splitting of the tin atomic position (Fig. 13a, c), which together with the partial ordering of the vacancies results in the formation of the superstructure of clathrate-I (vide supra).

Substitution of part of tin atoms with a heterometal was shown to be an effective way of producing vacancy-free cationic clathrates-I. All such clathrates contain 22 pnicogen atoms in the framework with the general formula $[Sn_{24-f}M_fPn_{22}]X_8$ (M = In, Zn, Cu; Pn = P, As; X = Br, I). Thus, the framework has $22 - 8 = 14$ extra electrons. Amount (*f*) of the substitution metal of the group *n* necessary for producing electron-balanced phase can be calculated as follows: $f = (22 - 8)/(14 - n)$. For the metals of groups 13, 12, and 11 such clathrates were successfully synthesized: $[Sn_{10}In_{14}P_{22}]I_8$ ($n = 13$, $f = 14$) [25], $[Sn_{17}Zn_7P_{22}]X_8$ ($n = 12$, $f = 7$) [51], and $[Sn_{19.3}Cu_{4.7}As_{22}]X_8$ ($n = 11, f = 14/3$) [50]. Depending on the nature of the heterometal, the clathrate phases exhibit different splitting of the 24*k* metal position (Fig. 13), which was elucidated by the combination of the single crystal X-ray diffraction and ^{119}Sn Mössbauer spectroscopy. For the vacancy-free phases such splitting appears due to the difference in the metal–metal bond length. The (In/Sn)–(In/Sn) bonds have similar length; consequently, no splitting of the 24*k* position was observed for $[Sn_{10}In_{14}P_{22}]I_8$. The Sn–Zn and Sn–Cu bonds are significantly shorter than the Sn–Sn ones, which ensures the splitting in $[Sn_{17}Zn_7P_{22}]X_8$ and $[Sn_{19.3}Cu_{4.7}As_{22}]X_8$ (Fig. 13a, d).

4.2.3 Thermoelectric Properties

The recent impetuous interest in clathrate compounds is caused by the prospects of their application as thermoelectric materials (TM) [16, 17, 28, 70, 116–123]. Such materials are able to transform the electrical energy directly into heat and *vice versa*. The figure-of-merit of a thermoelectric material is a dimensionless coefficient *ZT*:

$$ZT = S^2T\sigma/\kappa,$$

where *S* is the thermopower, *T* is the absolute temperature, σ is the electrical conductivity of the material, and κ is its thermal conductivity.

Clathrates were considered promising materials for TM cooling, i.e., for conversion of electrical power into active cooling at the temperatures below 300 K [16, 17]. It was predicted that thermoelectric materials with $ZT > 2$ can substitute for conventional cooling systems. Specific features of the clathrate crystal structures are responsible for their anomalously low thermal conductivity comparable with that of glasses. For Zintl phases, the electronic contribution to the thermal conductivity is typically low, not exceeding 10%. The rest of the thermal conductivity lies within the lattice part, which is usually calculated from the total thermal conductivity by subtracting the electronic contribution obtained using the Wiedermann–Franz law: $\kappa_e = L_0 T/\rho$, where the ideal Lorentz number is taken as $L_0 = 1/3(\pi k_B/e)^2$.

Low lattice thermal conductivity of Zintl clathrates stems from the interaction of spatially segregated host and guest substructures. Rattling of the guest atoms trapped inside the framework cages is believed to stimulate scattering of heat-carrying phonons. Recently, a new scenario for having abnormally low values of thermal conductivity was suggested, which involves the avoided crossing of the rattler modes and branches of acoustic phonons leading to low phonon group velocities [124]. The thermal conductivity can be tuned by varying the guest-atom size and mass. Typically, the increasing of the guest mass, host–guest mismatch, and disorder in the guest sublattice causes the decrease of the thermal conductivity [121]. For the majority of clathrates, the thermal conductivity at room temperature is lower than 1 Wm^{-1} K^{-1} and for certain compounds, e.g., $[Sn_{24}P_{19.2}\square_{2.8}]Br_2I_6$, it drops below 0.5 Wm^{-1} K^{-1} (Fig. 14) [59, 91]. The investigation of the physical grounds for abnormally low thermal conductivity with relation to the crystal and electronic structure of various Zintl clathrates has been one of the key issues in the thermoelectric clathrates research. Readers are referred to the following comprehensive works and references therein [17, 28, 70, 73, 126–129].

The $S^2\sigma$ term, called the power factor, depends on the band structure of the compound. Clathrates provide the unique possibility of band structure tuning by means of vacancy formation and heteroatomic substitutions. Thus, the thermal and charge carriers transport properties of Zintl clathrates may be optimized to some

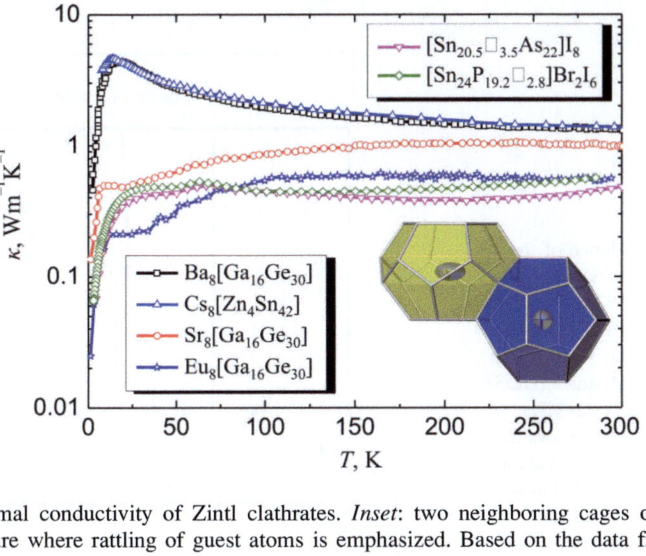

Fig. 14 Thermal conductivity of Zintl clathrates. *Inset*: two neighboring cages of clathrate-I crystal structure where rattling of guest atoms is emphasized. Based on the data from [59, 91, 111, 125]

extent independently. As a result of the optimization attempts, Zintl clathrates with the electrical conductivity varying from several tens to 10^5 S m^{-1} were synthesized, while the thermopower of n-type conductors can reach –800 μV K^{-1} at room temperature [121]. Nonetheless, the best thermoelectric figure-of-merit at 400 K was merely 0.4 for $Eu_8[Ga_{16-x}Ge_{30+x}]$ [130].

Higher *ZT* is not achieved at room temperature for type I clathrates mainly due to the fact that the thermal and charge carriers transport properties of Zintl clathrates are still coupled, i.e., changes in the local atomic surrounding and/or composition of the guest sublattice may significantly affect the band structure and charge carriers concentrations due to small adjustments in the crystal structure. This was demonstrated for the solid solution $[Sn_{24}P_{19.2}\square_{2.8}]Br_xI_{8-x}$ [91, 115]. The substitution of bromine for iodine results in the chemical compression of the clathrate framework due to smaller radius of bromine. At the first approximation, the guest substitution should not affect the electrical conductivity since the band gap and electrical conductivity are defined by the framework rather than by the X^- guests. However, experimental studies show that the band gap of $[Sn_{24}P_{19.2}\square_{2.8}]Br_xI_{8-x}$ depends linearly on the bromine content, x (Fig. 15). Quantum-chemical calculations reveal that the states just below the Fermi level in $[Sn_{24}P_{19.2}\square_{2.8}]I_8$ mainly originate

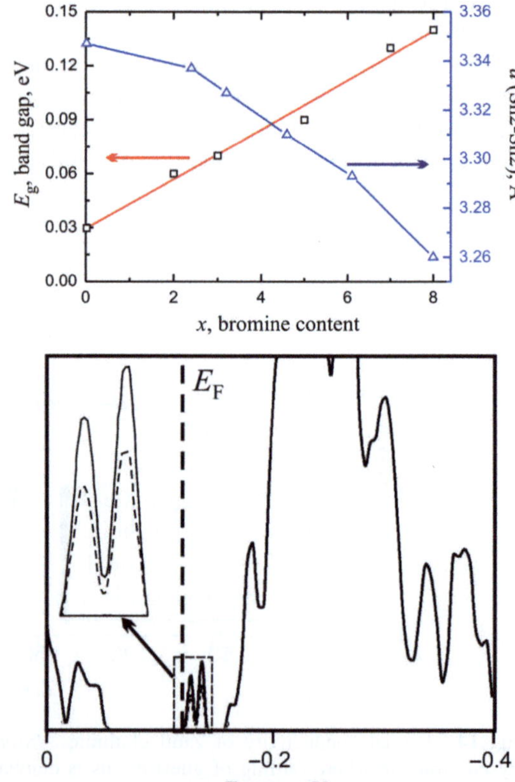

Fig. 15 Dependence of the band gap (*squares*) and Sn2–Sn2 distance (*triangles*) of $[Sn_{24}P_{19.2}\square_{2.8}]Br_xI_{8-x}$ on the bromine content, x (*top*); part of the density-of-states (DOS) plot for $[Sn_{24}P_{19.2}\square_{2.8}]I_8$ (*bottom*), the *inset* shows the states at the top of the valence band where the contribution of three-coordinated tin atoms, Sn2, is shown by the *dashed line* [25, 91, 115]

from the three-coordinated tin atoms surrounding the vacancy, and namely these atoms contribute substantially to the transport of the charge carriers (Fig. 15) [25]. Careful analysis of the structural data shows that the chemical compression of the host framework upon substitution in the guest sublattice is realized through the shortening of the distances between such three-coordinated tin atoms. The dependence of the distances between three-coordinated tin atoms upon the Br content is almost linear, correlating well with the linear dependence of the band gap upon the Br content in the solid solution $[Sn_{24}P_{19.2}\square_{2.8}]Br_xI_{8-x}$ [91].

Another aspect of TM materials is converting heat into electric energy. Considerably higher values of ZT were reported for some clathrates at $T \leq 1{,}000$ K compared to the room temperature values. The record holder is the Czochralski-grown single crystal of $Ba_8Ga_{16}Ge_{30}$ with ZT value of 1.35 at 900 K [131]. Pressed or sintered powder samples exhibit significantly lower ZT values at the same temperature [132–135]. The fine tuning of the clathrate composition with the aim of further increasing of the power factor resulted in achieving the ZT value of 1.2 at 1,000 K for the pressed powder sample with the composition $Ba_8[Ni_{0.31}Zn_{0.52}$-$Ga_{13.06}Ge_{32.2}]$ [135]. Although the Zintl clathrates might be useful in mid-temperature applications, as a rule they are not applicable at high temperatures since such uses require high stability of the materials at temperatures ~1,000 K. Most of the discussed Ge- and Sn-based clathrates are not stable in air at such conditions. Recently some of the Si-based clathrates started to be considered for high-temperature TM applications. Classical anionic clathrates $Ba_8[Al_xSi_{46-x}]$ and $EuBa_7[Al_xSi_{46-x}]$ display relatively high ZT values, up to 0.34 at 1,100 K [136, 137]. Inverse cationic clathrate-I $[Si_{46-x}P_x]Te_{8-\Delta}$ was shown to be stable at higher temperatures and, most importantly, stable in air at least below 1,290 K [48, 72]. Such extreme air stability is ensured by the formation of the silica nanolayer on the clathrate surface. The advantage of the Si–P–Te clathrates is the possibility to tune their composition and, correspondingly, transport properties by varying the Si/P ratio. Preliminary investigations of the TM properties of the type I Si–P–Te clathrate showed ZT of 0.36 at 1,100 K for nonoptimized compositions, which is comparable with the anionic analogs [138].

4.3 Clathrate-II

A recent review of Beekman and Nolas [70] surveys the synthetic methods, crystal and electronic structure, and transport properties of the type II clathrates. Therefore, we address to that review to those readers who are interested in details of the electronic structure and properties. In this section, we will summarize the important features concerning the electron count, bonding, conductivity, and thermoelectric properties of clathrates-II.

The type II clathrates are unique among all Zintl clathrates in that the violation of the Zintl rule is more norm than exception. These clathrates typically exhibit framework atoms of one chemical nature. With all framework atoms being four-bonded, only the guest-free clathrates made of the group 14 element will be charge-balanced. Indeed, such a form of germanium, Ge_{136}, was recently discovered [19]. According to the optical, magnetic, and resistivity measurements, this compound is diamagnetic and a direct band semiconductor ($E_g = 0.6$ eV), which agrees with the expected Zintl-like behavior. The silicon analog of Ge_{136} has never been obtained, and the name Si_{136} is used to underline that Na_xSi_{136} has the marginal sodium content ($x \rightarrow 0$) [18, 139, 140]. The latter clathrate also shows the nonmetallic type of electrical conductivity; however, it is noticeable that the clathrate form of germanium possesses by 0.1 eV narrower band gap than the diamond-like germanium; it is the other way around for silicon where the expanded band gap reaches 1.9 eV compared to 1.1 eV for α-Si [139]. Given only insignificant amount of sodium in the clathrate form of silicon, the interaction of the silicon sp^3 hybrids with the orbitals of alkali metal can be disregarded; therefore, the expansion of the band gap is the intrinsic property of Si_{136}. Calculations suggest [141] that the band tuning is possible in a wide energy range for the hypothetical solid solution $Si_{136-x}Ge_x$, and that only with high germanium content this clathrate should be a direct band semiconductor.

A maximum of 24 alkali metal atoms can be inserted in the cages of Si_{136}, which is accompanied by the ionization of the guests. Since there is no mechanism for the charge compensation, the resulting clathrates should exhibit metallic properties. Band-structure calculations show that Si_{136}, Na_8Si_{136}, and $Na_{24}S_{136}$ all have rather wide band gap, with all bonding states below the gap being filled [142, 143]. The antibonding (originally Si–Si) states above the gap accommodate electrons coming from the alkali metal atoms, up to 24 electrons for the clathrate with the maximal sodium content $Na_{24}Si_{136}$. The Na atoms are not fully ionized; the 3*s* Na orbitals strongly contribute to the states just above the Fermi level, modifying the bonding picture and electronic properties. At lower sodium concentration ($x \leq 8$), the interpretation of the results of band structure calculation differ [142, 144–146]; however, in general, the metal-to-semiconductor transition is expected.

The prediction that the A_xSi_{136} clathrate should exhibit metallic properties is experimentally confirmed for such compounds as $Cs_8Na_{16}Si_{136}$ [68, 147], $Ba_8Na_{16}Si_{136}$ [148], and $Na_{22}Si_{136}$ [149]. The former clathrate not only shows the monotonic increase of resistivity with temperature but also exhibits temperature-independent Pauli-like magnetic susceptibility, therefore, behaving as a normal metal. Analogous Ge-based clathrates also display metallic properties. For instance, both $Cs_8Na_{16}Ge_{136}$ and $Cs_8Na_{16}Cu_5Ge_{131}$ [150] are shown to have the temperature dependence of electrical conductivity typical for metals. The introduction of copper into the framework results in the ten times larger electrical resistivity compared to the parent compounds, which is attributed to lower concentration of available charge carriers. Decreasing the sodium content in Na_xSi_{136} first leads to the increase in the electrical resistivity, and then, between $x = 7$ and 12 [70, 151], the clathrate becomes a semiconductor and the room-temperature resistivity reaches values of 10^3–10^7 Ω m depending on the sodium concentration [152]. The availability of the metal-to-insulator transition together with the observation of superconductivity in type I clathrates motivated the low-temperature measurements of conductivity and magnetic susceptibility of type II clathrates; however, none of the investigated compounds showed superconductivity at least above 2 K [68, 148, 153].

NMR spectroscopy data provide further information on the electronic structure of the type II clathrates. Various ^{23}Na and ^{29}Si NMR measurements show very large Knight shifts with a strong temperature dependence atypical for conventional metals [139, 154–156]. This indicates the low degree of the guest atom ionization and the interaction of their valence *s* electrons with the silicon nuclei. Somewhat different picture is observed in the ^{87}Rb and ^{133}Cs NMR spectra. Although the similar temperature dependence remains in these spectra, the Knight shift is lower than that reported for metallic rubidium and cesium [154, 157], which is probably associated with the degree of the charge transfer from the guest to the framework conduction band. Interestingly, a complex feature observed in the ^{133}Cs spectrum of Cs_8Ge_{136} between 358 and 378 K is interpreted by Gryko et al. as the evidence of the Cs–Cs dimers formation, which have never been observed by other means [157].

Thermoelectric properties of the type II clathrates have recently become the point of great interest. The above-mentioned review [70] pays particular attention to the transport and thermal properties of these compounds. They point at the fact that all guest atoms occupying the 28-vertex cages in type II silicon and germanium clathrates display much larger anisotropic displacement parameters than the framework atoms [147]. Such a rattling of the guest atom is frequently used for explaining the amorphous-like thermal conductivity of clathrates. Although such approach is useful, it should be taken into account that the type II clathrates are divided into two groups of compounds according to their electrical conductivity; those that are metals have a considerable contribution of electronic component to the total thermal conductivity, whereas thermal conductivity of those that are semiconductors is almost entirely of the lattice nature. Thermal conductivity of Na_xSi_{136} ($x = 1$ and 8) has only negligible electronic contribution and is rather low, and the room temperature value for Na_8Si_{136} being slightly over 1 W m^{-1} K^{-1} [152]. Thermal conductivity of almost guest-free "Si_{136}" is also rather low (2.5 W m^{-1} K^{-1} at room temperature), which is attributed to the great effect of a large number of atoms in the crystal structure on the phonon transport [158]. This might have indicated that the rattling of guest atoms is not necessary to achieve the low thermal conductivity, but as soon as the measurements were performed on the pellets with the 70% density, the influence of the porosity prevents making conclusions about the actual thermal conductivity of Si_{136}. The insertion of sodium atoms ensures the decrease in thermal conductivity even compared to porous Si_{136}, once the semiconducting behavior is preserved. At higher sodium concentrations, the substantial increase in thermal conductivity is observed [149]. It is attributed to the large electronic contribution, which in $Na_{24}Si_{136}$ is comparable with the total thermal conductivity reaching high values of 24 W m^{-1} K^{-1} at 300 K [153]. It is also important to note that the thermopower has a negative sign and is quite low by absolute value for the type II clathrates [149, 150, 153], which is normal for metals with a high density of electrons. Altogether, this leads to low values of the thermoelectric figure-of-merit. However, the possibility of tuning the properties of type II clathrates by various substitutions raises hopes of finding new thermoelectric materials [150].

4.4 Clathrate-VIII

Although only four representatives of this type have been reported to the date, the literature reporting the electronic structure and properties of these compounds comprises more than a dozen research papers. Type VIII clathrates have the same stoichiometry as type I counterparts; moreover, in certain cases, type I and type VIII clathrates have the same chemical composition. For instance, the low-temperature phase α-$Eu_8Ga_{16}Ge_{30}$ is the type VIII clathrate; it undergoes the phase transition at 970 K into β-$Eu_8Ga_{16}Ge_{30}$, which is the type I clathrate [73, 126]. Of course, similar electron count applies and both phases comply with the Zintl scheme, $(Eu^{2+})_8(Ga^-)_{16}(Ge^0)_{30}$. The bandstructure calculations performed for the ideal composition $Ba_8Ga_{16}Sn_{30}$ [159] and hypothetical $A_8Ga_{16}Ge_{30}$ (A = Sr, Ba, Eu) [160] predict these compounds to behave as normal low-gap semiconductors. At the same time, a narrow band in the middle of the gap is found in the calculated band structure for $Ba_8Ga_{16}Sn_{30}$, which is attributed to the hybridization of the anti-bonding states of the Ga/Sn framework with the orbitals of the guest atoms. This band is associated with the observed superconductivity at 4 K. However, the authors point at the closeness of T_C to that of β-Sn but notice that the volume of the superconducting phase (20%) is much greater than the possible admixture of gray tin [159].

In fact, clathrate-VIII in the Eu–Ga–Ge system exhibits a homogeneity range and should be formulated as $Eu_8Ga_{16-x}Ge_{30+x}$ where $0.28 \leq x \leq 0.48$ at 923 K. This means that the proper Zintl composition cannot be achieved under given conditions and the properties depend on the composition [126, 130]. For the entire compositional range, $Eu_8Ga_{16-x}Ge_{30+x}$ displays a slight increase in the electrical resistivity at low T, and then $\rho(T)$ is linear above 100 K. Together with rather low concentration of charge carriers and rather high absolute values of the Seebeck coefficient, up to $-140\ \mu V\ K^{-1}$, this suggests the semimetallic behavior of α-$Eu_8Ga_{16}Ge_{30}$. Similar semimetallic properties are reported for $Sr_8Al_xGa_{16-x}Si_{30}$ ($8 < x < 13$) [75, 161] and $Sr_8Al_xGa_yGe_{46-x-y}$ ($6 \leq x \leq 7$, $10 \leq y \leq 11$) [76]. Based on the observed properties, these phases were called the "n-type degenerate semiconductors." In contrast, $Ba_8Ga_{16}Sn_{30}$ can be synthesized both as the gallium-rich and gallium-depleted compound, which leads to different properties depending on whether electrons or holes are the dominant charge carriers [162–164]. The samples grown from the tin flux are the n-type semiconductors with rather high Seebeck coefficient reaching $-243\ \mu V\ K^{-1}$ at 550 K, which is attributed to low charge carrier concentration [162]. The samples prepared using the gallium flux method have holes as dominate charge carries and display lower values of thermopower, not exceeding $+80\ \mu V\ K^{-1}$ at room temperature [163].

In type VIII clathrates, guest atoms rattle inside the oversized cages of irregular symmetry (Fig. 10). As a result, the thermal conductivity of these compounds is rather low. The electronic contribution to the total thermal conductivity is negligible below 100 K, and in this region the observed values can be regarded as pure lattice thermal conductivity. It is shown that the lowest values are found for

$Ba_8Ga_{16}Sn_{30}$, where the thermal conductivity of the *p*-type samples is glass-like and lower than 1 W m^{-1} K^{-1} [164]. A change in the majority carriers' type from holes to electrons leads to the transition from the glass-like to crystalline-like thermal conductivity with $\kappa_{100\ K} = 1.1$ W m^{-1} K^{-1}, which is attributed to the better coupling of the guest vibration modes with the phonons of the *p*-type framework. Low values of thermal conductivity together with rather high electrical conductivity and thermopower give rise to promising values of the thermoelectric figure-of-merit. For instance, the *n*-type $Sr_8Al_{6.3}Ga_{10.3}Ge_{29.4}$ displays $ZT = 0.56$ at 800 K [76].

α-$Eu_8Ga_{16}Ge_{30}$ is a rare example of the clathrate with the full population of the guest positions by a magnetic cation. Europium has the ground state configuration $4f^7$ expected for the Eu^{2+} cation, which is confirmed by the X-ray absorption spectroscopy (XAS) and magnetization measurements. At high temperature it is a Curie–Weiss paramagnet, and at 10 K the spins order ferromagnetically, probably, due to the Rudermann–Kittel–Kasuya–Yosida interaction [126]. α-$Eu_8Ga_{16}Ge_{30}$ undergoes a second-order phase transition with the sharp change of magnetization at the transition temperature. This is associated with the observed giant magneto-caloric effect with the low-filed magnetic entropy change of about 11.3 J kg^{-1} K^{-1} at 13 K [165, 166]. The estimated cooling capacity of α-$Eu_8Ga_{16}Ge_{30}$ is 87 J kg^{-1}, making this clathrate a good candidate for low-temperature magnetic refrigeration.

4.5 Clathrate-IX

Eight of the 25 framework atoms of the type IX clathrates are three-bonded (3*b*), which makes the electronic structure of clathrate-IX different from that of other clathrates. This difference shows up in the apparent localization of the electron pair on the 3*b* atom. These 3*b* atoms are treated as five-electron entities within the frames of the Zintl rule in such a way that three electrons are used to form two-center, two-electron bonds and two other electrons are localized in a form of the lone pair. For example, the distribution of the formal charges in $Ba_6Ge_{21}In_4$ [83] leads to the charge-balanced picture $(Ba^{2+})_6[(4b)Ge^0]_{13}[(4b)In^-]_4[(3b)Ge^-]_8$. Another way to balance the charge is to remove 1 of 25 framework atoms, which is observed in Ba_6Ge_{24} [82]. For such compounds, the expected semiconducting properties are indeed observed. It should be noted that Ba_6Ge_{25} with all framework positions occupied is also reported; moreover, it is argued that only Ba_6Ge_{25} exists as the equilibrium phase according to the binary Ba–Ge phase diagram [87].

Fulfilling the Zintl rule is not a common picture for the type IX clathrates. There are two groups of compounds that violate this rule in two different ways, being either electron-excessive or electron-deficient. The example of the electron-excessive clathrate-IX is provided by the above-mentioned compound Ba_6Ge_{25}, which has four extra electrons per formula. Ba_6Ge_{25} exhibits a structural transformation between 170 and 200 K associated with the rearrangement of the certain part of germanium atoms accompanied by the shift of one sort of barium atoms from the ideal position. This leads to the partial Ge–Ge bond breaking and appearance of new 3*b* Ge atoms [87]. As a consequence, the number of excess electrons reduces from 4 per formula to approximately 2.25 per formula. Upon this transformation, the compound retains its metallic properties; however, both the electrical resistivity and thermopower show a steep increase upon cooling, while the magnetic susceptibility decreases. At the same time, this structural transformation imposes no effect on the thermal conductivity, mainly because the electronic contribution to the total thermal conductivity is very low [167]. The temperature dependence of the specific heat displays a positive deviation from the Dulong–Petit law slightly above 200 K, which is attributed to the ability of the guest Ba atoms to migrate through the labyrinth within the guest substructure [88] until the structural transformation closes the pathways, forcing these guests freeze at their "proper" positions [87, 168]. The freezing of the barium guests is confirmed by the XPS measurements, which evidence a large reduction in the density of states at the Fermi level before and after the structural transformation as well as a substantial reduction of hybridization between the atomic orbitals of barium and germanium below the transition temperature [168]. The bandstructure calculations [169, 170] suggest that the driving force of the structural transformation is the shift of a part of barium atoms from their ideal positions to remove structural instability manifested by the peak in the density of states at the Fermi level. This band analog of the Jahn–Teller distortion is accomplished in expense of energy, and lowering the temperature ensures the stability of the ordered structure. It is interesting to note that the partial replacement of Ba by Na or Eu suppresses the transition [167].

At low temperatures, Ba_6Ge_{25} and $Ba_4Na_2Ge_{25}$ become superconductors, with the transition temperatures lying below 1 K under normal pressure [15] but rising up to 3.85 K upon applying the external pressure of 2.8 GPa [171]. Superconductivity with $T_C = 1.55$ was also observed for Ba_6Si_{25} [172]. Data of various physical methods show that the rattling of the guest atoms is associated with the electron–phonon coupling, which is weaker than in clathrate-I Ba_8Si_{46} due to a large amount of three-bonded framework atoms; therefore, the critical temperature is lower for the type IX clathrate than for its type I analog [173].

Partial substitution of Ga for Ge in Ba_6Ge_{25} is the other way to reduce the number of excess charge carriers. It is shown [174] that up to 3.75 Ge atoms per formula can be replaced by Ga without any structural changes, which leads to the formula $Ba_6Ga_{3.75}Ge_{21.25}$ (or $Ba_{24}Ga_{15}Ge_{85}$, considering the unit cell content). At the limit of the substitution, the concentration of the excess electrons decreases down to 0.25 per formula. Despite such a low concentration, the clathrate retains its metallic conductivity. At the same time, the alloying with gallium increases the thermopower, which reaches the value of –150 μV K^{-1} at 943 K. Also, the Ga-substituted clathrate features a decrease in the thermal conductivity both due to alloying and optimizing the host–guest mismatch. Altogether, this leads to high values of the dimensionless thermoelectric figure-of-merit, *ZT*, reaching 1.25 at 943 K [174]. Similar Al- and In-substituted clathrates were also reported, the substitution limit being the same for indium and slightly lower for aluminum, reaching the maximum expressed by the formula $Ba_6Al_3Ge_{22}$ ($\equiv Ba_{24}Al_{12}Ge_{88}$)

[175, 176]. These compounds display similar trends in thermoelectric properties, though the highest *ZT* values observed for the maximal content of the substituting element are considerably lower. At the same time, the difference in the scattering power of Al or In compared to Ge enabled determining the site preference for the substituting atoms, which confirmed that both aluminum and indium atoms occupy positions with the tetrahedral coordination.

The observation of the intriguing thermoelectric properties for $Ba_6Ga_{3.75}Ge_{21.25}$ deserves more discussion. It is evident that reducing the number of excess charge carriers from 4 to 0.25 per formula increases the figure-of-merit by more than two orders of magnitude. At the same time, the limiting composition $Ba_6Ga_{3.75}Ge_{21.25}$ provides only a marginal deviation from the Zintl count for electron-precise compounds. Although this clathrate retains the metallic type of conductivity, it already demonstrates the reasonably high value of thermopower, which is by absolute value much greater than typical values for metals. Probably, this opens the proper direction for searching for efficient thermoelectric materials; that is, they should exhibit only a slight deviation from the Zintl scheme, which leads them from low-gap semiconductors toward bad metals or heavily doped semiconductors.

K_6Sn_{25} [85] provides examples of electron-deficient clathrate-IX, which can be rationalized as $(K^+)_6[(4b)Sn^0]_{17}[(3b)Sn^-]_8$. Such a formula indicates that this clathrate is two electrons short to be charge-balanced. Substituting two Bi for Sn produces the electron-precise compound $K_6Sn_{23}Bi_2$ [85, 177]. In the latter phase, bismuth atoms occupy the 3*b* positions, and thus only six 3*b* tin atoms remain in the framework. This concentration is exactly required to compensate for the positive charge of six K^+ cations according to the formula $(K^+)_6[(4b)Sn^0]_{17}[(3b)Sn^-]_6[(3b)Bi^0]_2$. The comparison of the electronic structures of K_6Sn_{25} and $K_6Sn_{23}Bi_2$ gives the explanation of the stability of electron-deficient clathrate-IX. Figure 16 schematically shows the band gap in the density of states below the Fermi level of the

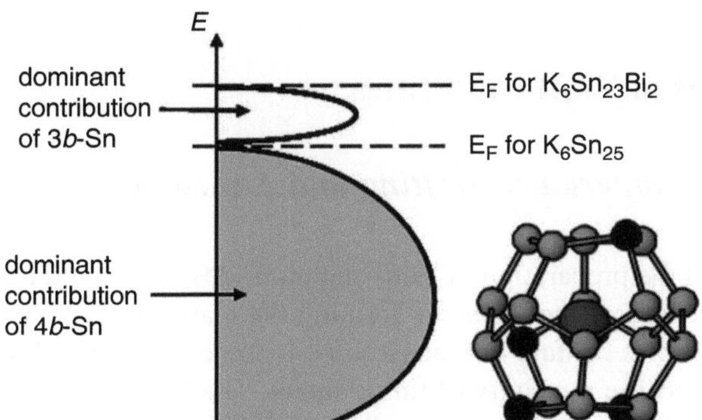

Fig. 16 Schematic presentation of the DOS plot for K_6Sn_{25} and $K_6Sn_{23}Bi_2$. A Sn_{20} cage with a guest K atom inside is shown, emphasizing the positions of 4*b*-Sn and 3*b*-Sn atoms. See [177, 178] for the detailed bandstructure presentation

Zintl compound $K_6Sn_{23}Bi_2$. This energy roughly corresponds to the Fermi level in K_6Sn_{25}. The region between the energies corresponding to the Fermi levels of different compounds is composed of bands coming predominantly from the lone pairs on tin atoms. These bands can accommodate eight electrons, which is exactly the difference between the number of electrons per unit cell ($Z = 4$) between $K_6Sn_{23}Bi_2$ and K_6Sn_{25}. Accounting for the features of its electronic structure, K_6Sn_{25} is classified as a "Zintl phase with a two-electron deficiency due to the interaction of the lone pairs" [178].

4.6 Other Clathrate Types

Zintl clathrates of types III and IV are not numerous and their electronic structures and properties are scarcely investigated. Type III anionic clathrates $A_{30}[Sn_{172-x}Na_y\square_z]$ (A = Cs/Rb) are diamagnetic [71]. The same is true for the cationic clathrate $[Si_{130}P_{42}]Te_{21}$ [72]. In the latter compound, electron balance is achieved by compensation of the negative charge of 21 guest tellurium atoms by 42 four-bonded phosphorus atoms of the vacancy-free framework. This compound exhibits the semiconductor type of electronic conductivity and displays relatively low thermal conductivity. This leads to rather high values of *ZT*, reaching 0.36 at 1,000 K for the nonoptimized compound. Coupled to the utmost chemical and thermal stability, this makes clathrates-III in the Si–P–Te system a promising candidate for the high-temperature thermoelectric power generation [138].

Very little is known about the electronic structure and properties of the sole clathrate-IV, $Li_{14.7}Mg_{36.8}Cu_{21.5}Ga_{66}$. According to the Extended Hückel tight-binding calculations, this compound should be metallic [23].

5 Synthesis of Zintl Clathrates

5.1 High-Temperature Melting and Annealing

The methods of the preparation of Zintl clathrates differ substantially depending on the chemical nature of the elements forming the clathrate framework. As soon as almost all clathrates contain one of Group 14 elements (Si, Ge, Sn), the synthetic procedure reflects the reactivity of this element. The least reactive silicon with the diamond-type crystal structure normally requires very high temperatures for its activation. For example, the preparation of $Ba_{8-x}Si_{46}$ required heating of the stoichiometric mixture of $BaSi_2$ and elemental silicon up to 1,573 K [14] and further annealing at 900 K for 24 h. It is easier to synthesize ternary and quaternary clathrate silicides by the stoichiometric method than their binary analogs [65, 132]. Very few papers report low-temperature synthesis of the Si-containing clathrates.

One of them is the synthesis of type II clathrates $A_8Na_{16}Si_{136}$ (A = Rb, Cs) by annealing the stoichiometric mixtures of elements at 823 K [21, 68]. Higher temperature (1,073 K) was necessary to complete the reaction of Ba, Eu, Al, and Si in 7 days, resulting in the phase-pure type I clathrate $EuBa_7Al_{13}Si_{33}$ [136]. Even higher temperatures were used to synthesize type I (1,375 K) and type III (1,425 K) clathrates in the Si–P–Te system [48, 72]. However, in most cases samples still need pretreatment for enhancing reactivity of the starting compounds. For this, arc melting of the sample is frequently used as, for example, in the case of $Sr_8[Ga_xSi_{46-x}]$ [179]. As a variation of this method, $K_7B_7Si_{39}$ was prepared in a two-step synthesis, with the first step being the arc melting of the mixture of boron and silicon; then, the resulting alloy was annealed with potassium at about 1,170 K to achieve phase-pure clathrate-I after subsequent washing of the product [29]. Arc melting is also used without further annealing and/or purification of the product. For instance, $Ba_8Cu_4Si_{42}$ was synthesized by arc melting of the elements under argon [180]. In some cases, due to the loss of volatile barium, the excess of this metal is introduced into the starting mixture [181]. Another way is the heating of the sample to very high temperatures, up to 1,700 K, in high-frequency furnace as in the case of $Ba_8[M_xSi_{46-x}]$, where M = Ni, Pd, Pt, Cu, Ag, or Au [32, 182].

Germanium is more reactive than silicon. Many clathrates can be prepared by direct annealing of the elements in vacuum or under argon pressure. For the preparation of Ba_6Ge_{25}, it was sufficient to anneal the stoichiometric mixture of Ba and Ge at 923 K for 12 days [87]. Higher temperature (1,313 K) was required for the preparation of $Eu_8Ga_{16-x}Ge_{30+x}$ starting from the elements [126]. The reaction of the pre-prepared Ba–Ge alloy with Zn was used to synthesize type I clathrates in the Ba–Zn–Ge system [183]. Arc melting followed by the rapid cooling of the sample is also frequently used for the preparation of the Ge-based clathrates, both binary and ternary [33, 34, 61, 184]. A two-step route involving the induction heating followed by the homogenizing annealing was used in the case of multicomponent systems; this procedure was successfully exploited for synthesizing $K_6Eu_2Ga_{10}Ge_{36}$, $K_6Eu_2Zn_5Ge_{41}$, and $K_6Eu_2Cd_5Ge_{41}$ [185].

Due to the low melting point of tin (505 K), Sn-based clathrates are generally prepared by the annealing of a stoichiometric mixture of the elements without pretreatment. The usual synthesis temperature falls in the range of 600–900 K [30, 56, 57, 60, 125]. Some tin-containing clathrates require quenching, for instance, $Ba_8Ga_{17.2}Sn_{28.8}$ [74].

Very few clathrates do not contain Group 14 element at all. Despite the variation in their chemical composition, they all can be prepared by direct reaction of the elements. This method was used for the preparation of $Cs_8M_{18}Sb_{28}$ at 823 K [37] and $Li_{14.7}Mg_{36.8}Cu_{21.5}Ga_{66}$ at 1,073 K [23]. For $Ba_8Cu_{16}P_{30}$ [36], the synthesis from CuCl, P, and excess Ba at 1,103 K was found to be the most effective, while the direct reaction of the elements resulted in the formation of $BaCu_2P_4$ instead of the clathrate phase.

Finally, it should be mentioned that high reaction temperatures imply that the choice of containers for the reactions is a very important point. Anionic clathrates are typically synthesized from more reactive precursors, viz. alkali metals; consequently, glassy carbon, tantalum, or alumina crucibles are required for the synthesis. Alternatively, starting compositions can be wrapped in Nb or Ta foils. Precursors for cationic clathrates are much less reactive toward quartz, and silica ampoules suffice.

5.2 Precursor Decomposition

Again, referring to the very low chemical activity of elemental silicon, we point to the necessity of elaborating efficient methods for the preparation of the silicon-based clathrates. One of such methods is the decomposition of precursors. Binary silicides MSi (M = Na, K, Rb) decompose to give $M_x[Si_{46}]$ or $M_x[Si_{136}]$ depending on the reaction conditions. Heating to 600–900 K in the absence of dynamic vacuum can afford either clathrate-I or clathrate-II. In dynamic vacuum, syntheses produce primarily type II clathrate compounds. Yamanaka et al. performed the detailed analysis of the decomposition of NaSi [186]. It was shown that under vacuum clathrate-I can be obtained only in the presence of sodium vapor, and above 793 K it decomposes, liberating silicon. Under similar conditions, clathrate-II is stable; it can be prepared between 633 and 713 K as a single phase with different sodium content, which seems to depend upon the decomposition temperature. For instance, the clathrates with the exact compositions Na_1Si_{136} and $Na_{22}Si_{136}$ were synthesized at 698 and 653 K, respectively [149, 152]. It was supposed that varying the synthetic conditions, including temperature, residual pressure, and duration of experiment, it would be possible to "degas" type II clathrates yielding a guest-free clathrate framework. Ammar et al. [18] used very high-vacuum conditions for the decomposition of NaSi precursor at various temperatures and thus achieved the composition $Na_{0.0058}Si_{136}$, but the complete removal of sodium was not possible even after annealing the sample with iodine in vacuum.

Higher temperatures are required to synthesize clathrates with guest potassium or rubidium atoms from their monosilicides. K_8Si_{46} and Rb_8Si_{46} were prepared from KSi and RbSi, respectively, at 939 K [103]. Ternary silicon clathrates can also be obtained by this method. $Na_{16}Ba_8Si_{136}$ was prepared by the decomposition of presynthesized Zintl phase Na_2BaSi_4 at 623 K [148]. It was noticed that at higher decomposition temperatures the target product was contaminated by the admixture of $Na_2Ba_6Si_{46}$. Some other silicon-based clathrates were prepared in a similar fashion, including $A_8Na_{16}Si_{136}$ (A = Rb, Cs) [154, 187]. Finally, there is an example of obtaining binary cationic clathrate $[Ge_{46-x}I_x]I_8$ ($x \approx 8/3$) by a decomposition of germanium iodide in an inert gas atmosphere [54].

5.3 High-Pressure Synthesis

High-pressure and high-temperature synthesis is widely used for the preparation of Zintl clathrates, especially those containing silicon. In some cases, this method leads to the preparation of clathrates which cannot otherwise be synthesized. Binary clathrates in the Si–Te system were prepared at 5 GPa and temperatures between 1,100 and 1,473 K [45]. Depending on the conditions, two polymorphs of $[Si_{38}Te_8]Te_8$ were isolated. Higher temperature (1,573 K) and lower pressure (3 GPa) were used to obtain superconductive $Ba_{7.7}Si_{46}$ [188]. Imai and Kikegawa directly observed the formation of clathrate-I Ba_8Si_{46} at the approximate pressure of 4.3 GPa [189]. By means of in situ X-ray diffraction experiments, they found that the formation of the target phase from $BaSi_2$ and Si in a *h*-BN capsule started at about 970 K and was complete at 1,070 K. Attempts to prepare similar compounds with strontium or calcium in place of barium were unsuccessful [189]; however, mixed-cation clathrates $Ba_{8-x}Ae_xSi_{46}$ (Ae = Sr, Ca) were synthesized at temperatures of 1,073–1,173 K and pressures of 3–5 GPa [190, 191]. Noticeably, the substitution limit for strontium is higher (75%) than for calcium (50%) reflecting the difference in the atomic volume. Similarly, ternary clathrates $Ba_{8-\Delta}Ge_xSi_{46-x}$ ($x \leq 23$) were synthesized from the mixtures of $BaSi_2$, $BaGe_2$, Si, and Ge at 1,073 K and 3 GPa [24]. There is no apparent optimal pressure range; however, it was shown that the formation of $Cs_{8-\Delta}Si_{46}$ from Cs_4Si_4 and α-Si was incomplete at the pressure of 2.1 GPa, while at 3.5 and 10 GPa the reaction resulted in almost phase-pure clathrates [192]. Importantly, the excess pressure may lead to the decomposition of clathrates, probably, because the clathrate structures are not dense and excessive compression leads to the structure collapse. For instance, Na_xSi_{136} ($x \rightarrow 0$) decomposes between 8 and 10 GPa with the formation of the β-Sn-type structure [193], in which silicon has a coordination number six.

A new variation of the high-pressure synthesis is provided by the spark-plasma-sintering (SPS) technique [194]. In this method, the preparation of the target compound is achieved by applying pressure to the sample together with the pulsed DC electric current (Fig. 17). The sample chamber can be either evacuated or filled with inert gas, typically Ar. This allows performing reactions at very high temperatures, which are only limited by the materials of a die. For example, graphite dies exhibit higher temperature stability up to 1,000°C while steel and tungsten carbide are more pressure resistant. Depending on the conditions and precursors, clathrates with different chemical composition were synthesized using the SPS technique. For instance, $Na_{24}S_{136}$ was prepared by processing Na_4Si_4 at 823–973 K and 100 MPa for about 3 h [195]. Lower pressure (40 MPa) and higher temperature (1,123–1,273 K) were used for the preparation of type I clathrates in the Si–P–Te, Ge–P–Te, and Sn(Ge)–Sb–I systems [49, 52, 196]. It was demonstrated [197] that the SPS synthesis can be used for the preparation of five-component compounds such as $Ba_6A_2AuGa_xSi_{45-x}$ (A = Sr, Eu; $x = 13, 14$). Another advantage of this technique is that it provides dense samples by treating presynthesized clathrates. A series of samples was obtained by SPS treatment of the preprepared $K_8Ga_xSn_{46-x}$, with x ranging from 6 to 8, at 753–823 K and 40 MPa [198].

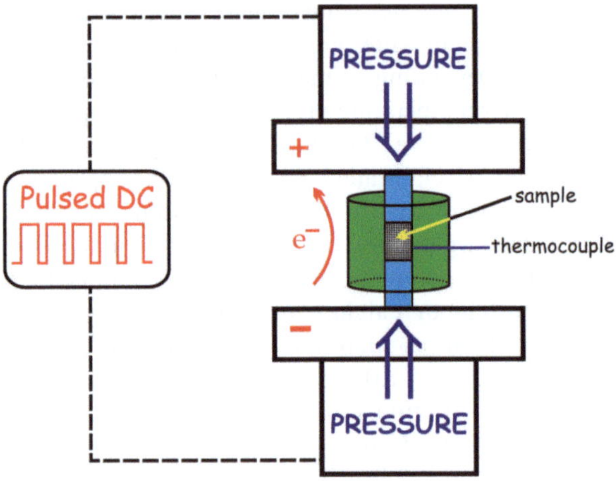

Fig. 17 Scheme of the SPS setup used for the synthesis of $Na_{24}Si_{136}$

5.4 Flux Synthesis

A variation of the flux synthesis used for the preparation of clathrates is the synthesis in molten metals. For many years, only tin flux was widely used; now the situation has changed, and other metals with low melting points such as gallium, aluminum, zinc, and even sodium are finding their application as a synthetic media [199]. For instance, the Al flux was used for the preparation of Al-containing clathrates [137, 200, 201]. Large single crystals, up to 60 mg of weight, were obtained by this technique. Ga flux was used for the preparation of single crystals of type I clathrates in the Ba–Ga–Ge system at 1,273 K, which is well below the melting point of germanium [113]. In a similar way, tin serves both as the flux and the reagent, for example, in the synthesis of the type IX clathrate $Rb_5Na_3Sn_{25}$ [86]. There is another technique that is used for the single crystal preparation; this is known as Czochralski pulling. Single crystals of both *n*- and *p*-type $Ba_8Ga_{16}Ge_{30}$ were prepared by pulling from the melt made of the stoichiometric mixture of the elements and heated above the melting point of germanium [26]. The highest *ZT* value for clathrates, 1.35 at 900 K, was achieved on the disk sample cut from Czochralski grown single crystal of $Ba_8Ga_{16}Ge_{30}$ [131].

5.5 Transport Reactions

Synthesis of the majority of cationic clathrates is accompanied by a gas transport reaction in response to the presence of volatile starting materials. Halogens are the most frequent guests in the cationic clathrates; however, due to their excessive chemical activity and because of the high vapor pressure, elemental halogens are only in exceptional cases used as the starting materials. Instead, less volatile halides are introduced into the starting mixture. For instance, the preparation of Sn-based clathrates always involves SnI_4 if iodine is the guest atom [25, 42, 51, 59, 115]. In these cases, the dissociation of SnI_4 starts at the reaction temperatures (750–875 K) liberating gaseous iodine, which concentration is low under such conditions and which serves as the transport agent. Other possible transport agents include TeI_4, $TeCl_4$, $HgBr_2$, and Hg_2Cl_2 [53, 115, 202]. The use of elemental halogens is limited; their choice is motivated by the absence of readily accessible sources of halogens. For instance, iodine serves as a reagent and simultaneously as a transport agent for the preparation of $Si_{40}P_6I_{6.5}$ [44]; however, the formation of numerous volatile byproducts prevents obtaining the phase-pure product. Of course, transport reactions are used for the preparation of single crystals of cationic clathrates in a temperature gradient, as it is shown for various compositions of the $Sn_{24}P_{19.3}I_{8-x}Br_x$ solid solutions [115].

5.6 *New Routes*

Several new preparation routes to various clathrates have been developed in recent years. One of them is the oxidation of precursors by gaseous HCl. This procedure leads to the preparation of $Na_2Ba_6Si_{46}$ from Na_2BaSi_4 [203]. Importantly, the side products of this reaction, $BaCl_2$ and NaCl, can be easily washed away by water. In a similar fashion, Na_4Si_4 and K_4Si_4 can be oxidized into type I clathrates at low temperatures by HCl or H_2O in high yields [204]. Recently, the variation of this method was shown to lead to hydrogen-containing clathrates [38, 39]. The reactions of NH_4Br with Na_4Si_4 or K_4Si_4 give $Na_{5.5}(H_2)_{2.15}Si_{46}$ or $K_7(H_2)_3Si_{46}$, respectively. Noticeably, the reaction temperatures differ for two hydrogen-containing clathrates, the sodium derivative requiring much lower temperature (573 K) than its potassium analog (923 K). Another unexpected route was designed for synthesizing $K_4Hg_{3+x}Ge_{43-x}$, which exploited the reaction of the Zintl phase K_4Ge_9 with the mixture of WO_3/HgO; it is worth noting that the direct interaction of K and Ge with Hg leads only to minor yields of the target clathrate [31]. Most excitingly, a guest-free germanium clathrate (Ge_{136}) was prepared by a mild oxidation of Na_4Ge_9 in ionic liquids at 573 K [19]. This method has led to the preparation of a previously unknown atmospheric polymorph of germanium, which was inaccessible by other means; however, the reaction is very whimsical, and a slight variation in the nature of either oxidant or precursor leads to different products. For instance, the same authors reported that when K_4Ge_9 was introduced into the same reaction followed by annealing the product, a new guest-containing germanium clathrate-II formulated as $K_{8.6}Ge_{136}$ formed instead of the guest-free (Ge_{136}) [67]. They also mentioned that clathrate-I K_8Ge_{44} could be prepared by this synthetic route [19].

6 Conclusion

Zintl clathrates comprise a vast family of cage compounds with a very rich chemistry. They crystallize into six structure types and involve 32 elements of the Periodic Table, which is 40% of all stable chemical elements. Their structures are not only esthetically beautiful; the arrangement of atoms and the host–guest interaction provide a plentiful field for modifying properties by various substitutions either in the clathrate framework, or in the guest substructure, or in both. Geometrical host–guest compatibility in clathrates is achieved through the formation of cages with different size and shape as well as by local distortions of the framework atoms environment. Electronic adjustment is realized by either introduction of vacancies/heteroatoms in the host framework or modifying band structure, which sometimes results in the transition from semiconductors to metallic phases. In most cases, the substitutions follow the predictions made by the application of the Zintl electron-count scheme. Even though this rule is not always adaptable, it helps to examine the correlations between the crystal structure, electronic structure, and various properties of Zintl clathrates. Several tools have already been exploited for the property tuning, including the adjustment of the host–guest mismatch aiming at optimization of the guest atomic rattling and the heterovalent substitutions changing the charge carrier concentration. Among the properties of Zintl clathrates, there are two that evoke much interest. One of them is superconductivity, which provides a rare example of the sp^3-hybridized superconductors. Another point of interest is the promising thermoelectric properties of Zintl clathrates. It is expected that new materials for waste heat recovery and direct solar thermal energy conversion will be constructed on the basis of properly tuned clathrates. Both properties rely upon the phonon structure of the clathrates. The electron–phonon coupling is responsible for superconductivity, whereas the low thermal conductivity is defined by localized vibrations of the guest atoms causing either the reduced phonon group velocity or the phonon scattering. In this sense, the "phonon engineering" is a hot topic in the clathrate chemistry and physics.

As new synthetic methods become available, the variation in the chemical composition and fine structural features of Zintl clathrate will be more and more accessible. Very likely, this will lead to new clathrates with unexpected properties and promising applications. Already now modern synthetic approaches have opened the gates to the clathrates, which would be even unpredictable a decade ago, such as, for instance, the guest-free Ge_{136} and hydrogen-containing $A_x(H_2)_ySi_{46}$ (A = Na, K). Probably, it is impossible to list all new properties that will be found in Zintl clathrates. The discovery of the giant magnetocaloric effect in α-$Eu_8Ga_{16}Ge_{30}$ with high cooling capacity at low temperatures is convincing us of the rich future of these fascinating compounds.

Acknowledgments The authors express their gratitude to Dr. J.V. Zaikina for critical discussion of various aspects of Zintl clathrates chemistry. Russian Foundation for Basic Research is acknowledged for the support.

References

1. Kasper JS, Hagenmuller P, Pouchard M (1965) Science 150:1713
2. Cros C, Pouchard M, Hagenmuller P (1965) C R Acad Sci Paris 260:4764
3. Wells A (1984) Structural inorganic chemistry, Chap. 15. Oxford University Press, London
4. Davy H (1811) Philos Trans R Soc London 101:155
5. Allen KW (1959) J Chem Soc 4131
6. Jeffrey GA (1984) In: Atwood JL, Davies JE, McNicol DD (eds) Hydrate inclusion compounds. Academic, London
7. Sloan ED Jr (1998) Clathrate hydrates of natural gases, 2nd edn. Marcel Decker, New York
8. Thomas S, Dawe RA (2003) Energy 28:1461
9. Koh CA, Sum AK, Sloan ED (2009) J Appl Phys 106:061101
10. Chatti I, Delahaye A, Fournaison L, Petitet J-P (2005) Energy Convers Manag 46:1333
11. Lee H, Lee JW, Kim DY, Park J, Seo YT, Zeng H, Moudrakovski IL, Ratcliffe CI, Ripmeester JA (2005) Nature 434:743
12. Hu YH, Ruckenstein E (2006) Angew Chem Int Ed 45:2011
13. Yamanaka S, Enishi E, Fukuoka H, Yasukawa M (2000) Inorg Chem 39:56
14. Fukuoka H, Kiyoto J, Yamanaka S (2004) J Phys Chem Solid 65:333
15. Grosche FM, Yuan HQ, Carrillo-Cabrera W, Paschen S, Langhammer C, Kromer F, Sparn G, Baenitz M, Grin Yu, Steglich F (2001) Phys Rev Lett 87:247003
16. Slack GA (1995) In: Rowe DM (ed) CRC Handbook of Thermoelectrics. Chemical Rubber, Boca Raton, FL
17. Nolas GS, Cohn JL, Slack GA, Schjuman SB (1998) Appl Phys Lett 73:178
18. Ammar A, Cros C, Pouchard M, Jaussaud N, Bassat JM, Villeneuve G, Duttine M, Ménétrier M, Reny E (2004) Solid State Sci 6:393
19. Guloy AM, Ramlau R, Tang Z, Schnelle W, Baitinger M, Grin Yu (2006) Nature 443:320
20. O'Keefe M, Hyde BG (1996) Crystal structures. 1. Patterns and symmetry. Mineral Society of America, Washington DC, p 53
21. Bobev S, Sevov SC (2000) J Solid State Chem 153:92
22. Kovnir KA, Shevelkov AV (2004) Russ Chem Rev 73:923
23. Lin Q, Corbett JD (2008) Inorg Chem 47:10825
24. Fukuoka H, Kiyoto J, Yamanaka S (2003) J Solid State Chem 173:237
25. Shatruk MM, Kovnir KA, Lindsjö M, Presniakov IA, Kloo LA, Shevelkov AV (2001) J Solid State Chem 161:233
26. Christensen M, Lock N, Overgaard J, Iversen BB (2006) J Am Chem Soc 128:15657
27. Christensen M, Iversen BB (2007) Chem Mater 19:4896
28. Christensen M, Johnsen S, Iversen BB (2010) Dalton Trans 39:978
29. Jung W, Lörincz J, Ramlau R, Borrmann H, Prots Yu, Haarmann F, Schnelle W, Burkhardt U, Baitinger M, Grin Yu (2007) Angew Chem Int Ed 46:6725
30. Kaltzoglou A, Ponou S, Fässler TF (2008) Eur J Inorg Chem 538
31. Kaltzoglou A, Ponou S, Fässler TF (2008) Eur J Inorg Chem 4507
32. Cordier G, Woll P (1991) J Less Common Met 169:291
33. Melnychenko-Koblyuk N, Grytsiv A, Rogl P, Rotter M, Lackner R, Bauer E, Fornasari L, Marabelli F, Giester G (2007) Phys Rev B 76:195124
34. Melnychenko-Koblyuk N, Grytsiv A, Rogl P, Rotter M, Bauer E, Durand G, Kaldarar H, Lackner R, Michor H, Royanian E, Koza M, Giester G (2007) Phys Rev B 76:144118
35. Nguyen LTK, Aydemir U, Baitinger M, Bauer E, Borrmann H, Burkhardt U, Custers J, Haghighirad A, Höfler R, Luther KD, Ritter F, Assmus W, Grin Yu, Paschen S (2010) Dalton Trans 39:1071
36. Dünner J, Mewis A (1991) Z Anorg Allg Chem 621:191
37. Liu Y, Wu L-M, Li L-H, Du S-W, Corbett JD, Chen L (2009) Angew Chem Int Ed 48:5305

38. Neiner D, Okamoto NL, Condron CL, Ramasse QM, Yu P, Browning ND, Kauzlarich SM (2007) J Am Chem Soc 129:13857
39. Neiner D, Okamoto NL, Yu P, Leonard S, Condron CL, Toney MF, Ramasse QN, Browning ND, Kauzlarich SM (2010) Inorg Chem 49:815
40. von Schnering HG, Menke H (1972) Angew Chem 84:30; Angew Chem Int Ed Engl 11:43
41. Menke H, von Schnering HG (1976) Z Anorg Allg Chem 424:108
42. Shatruk MM, Kovnir KA, Shevelkov AV, Presniakov IA, Popovkin BA (1999) Inorg Chem 38:3455
43. Reny E, Yamanaka S, Cros C, Pouchard M (2000) Chem Commun 24:2505
44. Kovnir KA, Uglov AN, Zaikina JV, Shevelkov AV (2004) Mendeleev Commun 4:135
45. Jaussaud N, Toulemonde P, Pouchard M, San Miguel A, Gravereau P, Pechev S, Goglio G, Cros C (2004) Solid State Sci 6:401
46. Jaussaud N, Pouchard M, Gravereau P, Pechev S, Goglio G, Cros C, San Miguel A, Toulemonde P (2005) Inorg Chem 44:2210
47. Zaikina JV, Kovnir KA, Schwarz U, Borrmann H, Shevelkov AV (2007) Z Kristallogr New Cryst Struct 222:177
48. Zaikina JV, Kovnir KA, Burkhardt U, Schnelle W, Haarmann F, Schwarz U, Grin Yu, Shevelkov AV (2009) Inorg Chem 48:3720
49. Kishimoto K, Akai K, Muroaka N, Koyanagi T, Matsuura M (2006) Appl Phys Lett 89:172106
50. Kovnir KA, Sobolev AV, Presniakov IA, Lebedev OI, Van Tendeloo G, Schnelle W, Grin Yu, Shevelkov AV (2005) Inorg Chem 44:8786
51. Kovnir KA, Shatruk MM, Reshetova LN, Presniakov IA, Dikarev EV, Baitinger M, Haarmann F, Schnelle W, Baenitz M, Grin Yu, Shevelkov AV (2005) Solid State Sci 7:957
52. Kishimoto K, Arimura S, Koyanagi T (2006) Appl Phys Lett 88:222115
53. Kovnir KA, Abramchuk NS, Zaikina JV, Baitinger M, Burkhardt U, Schnelle W, Olenev AV, Lebedev OI, Van Tendeloo G, Dikarev EV, Shevelkov AV (2006) Z Kristallogr 221:527
54. Nesper R, Curda J, von Schnering HG (1986) Angew Chem Int Ed Engl 25:350
55. Carrillo-Cabrera W, Budnyk S, Prots Yu, Grin Yu (2004) Z Anorg Allg Chem 630:2267
56. Dubois F, Fässler TF (2005) J Am Chem Soc 127:3264
57. Kaltzoglou A, Hoffmann SD, Fässler TF (2007) Eur J Inorg Chem 4162
58. Kaltzoglou A, Fässler TF, Christensen M, Johnsen S, Iversen B, Presniakov I, Sobolev A, Shevelkov A (2008) J Mater Chem 18:5630
59. Zaikina JV, Kovnir KA, Sobolev AV, Presniakov IA, Prots Yu, Baitinger M, Schnelle W, Olenev AV, Lebedev OI, Van Tendeloo G, Grin Yu, Shevelkov AV (2007) Chem Eur J 13:5090
60. Kaltzoglou A, Fässler TF, Gold C, Scheidt EW, Scherer W, Kume T, Shimizu H (2009) J Solid State Chem 182:2924
61. Okamoto NL, Oh MW, Nishii T, Tanaka K, Inui H (2006) J Appl Phys 99:033513
62. Pani M, Palenzona A (2008) J Alloys Comp 462:L9
63. Aydemir U, Candolfi C, Borrmann H, Baitinger M, Ormeci A, Carrillo-Cabrera W, Chubilleau C, Lenoir B, Dauscher A, Oeschler N, Steglich F, Grin Yu (2010) Dalton Trans 39:1078
64. Kröner R, Peters K, von Schnering HG, Nesper R (1998) Z Kristallogr New Cryst Struct 213:664
65. Beekman M, Wong-Ng W, Kaduk JA, Shapiro A, Nolas GS (2007) J Solid State Chem 180:1076
66. Mansour AN, Beekman M, Wong-Ng W, Nolas GS (2009) J Solid State Chem 182:107
67. Guloy AM, Tang Z, Ramlau R, Böhme B, Baitinger M, Grin Yu (2009) Eur J Inorg Chem 17: 2455
68. Bobev S, Sevov SC (1999) J Am Chem Soc 121:3795
69. Beekman M, Nenghabi EN, Biswas K, Myles CW, Baitinger M, Grin Yu, Nolas GS (2010) Inorg Chem 49:5338
70. Beekman M, Nolas GS (2008) J Mater Chem 18:842

71. Bobev S, Sevov SC (2001) J Am Chem Soc 123:3389
72. Zaikina JV, Kovnir KA, Haarmann F, Schnelle W, Burkhardt U, Borrmann H, Schwarz U, Grin Yu, Shevelkov AV (2008) Chem Eur J 14:5414
73. Paschen S, Carrillo-Cabrera W, Bentien A, Tran VH, Baenitz M, Grin Yu, Steglich F (2001) Phys Rev B 64:214404
74. Carrillo-Cabrera W, Cardoso Gil R, Tran VH, Grin Yu (2002) Z Kristallogr New Cryst Struct 217:181
75. Kishimoto K, Ikeda N, Akai K, Koyanagi T (2008) Appl Phys Exp 1:031201
76. Sasaki Y, Kishimoto K, Koyanagi T, Asada H, Akai K (2009) J Appl Phys 105:073702
77. Leoni S, Carrillo-Cabrera W, Grin Yu (2003) J Alloys Compd 350:113
78. Fukuoka H, Ueno K, Yamanaka S (2000) J Organomet Chem 611:543
79. Fukuoka H, Iwai K, Yamanaka S, Abe H, Yoza K, Häming L (2000) J Solid State Chem 151:117
80. Carrillo-Cabera W, Curda J, von Schnering HG, Paschen S, Grin Yu (2000) Z Kristallogr New Cryst Struct 215:207
81. Carrillo-Cabera W, Curda J, Peters K, Paschen S, Grin Yu, von Schnering HG (2001) Z Kristallogr New Cryst Struct 216:183
82. Kim SJ, Hu S, Uher C, Hogan T, Huang B, Corbett J, Kanatzidis MG (2000) J Solid State Chem 153:321
83. von Schnering HG, Kröner R, Carrillo-Cabera W, Peters K, Nesper R (1998) Z Kristallogr New Cryst Struct 213:665
84. Zhao JT, Corbett JD (1994) Inorg Chem 33:5721
85. Fässler TF, Kronseder C (1998) Z Anorg Allg Chem 624:561
86. Bobev S, Sevov SC (2000) Inorg Chem 39:5930
87. Carrillo-Cabrera W, Borrmann H, Paschen S, Baenitz M, Steglich F, Grin Yu (2005) J Solid State Chem 178:715
88. von Schnering HG, Zürn A, Chang JH, Baitinger M, Grin Yu (2007) Z Anorg Allg Chem 633:1147
89. Zintl E (1939) Angew Chem 52:1
90. Miller GJ (1996) In: Kauzlarich SM (ed) Chemistry, structure, and bonding of Zintl phases and ions. VCH, New York, and references therein
91. Zaikina JV, Schnelle W, Kovnir KA, Olenev AV, Grin Yu, Shevelkov AV (2007) Solid State Sci 9:664
92. Yamanaka S (2010) Dalton Trans 39:1901
93. Kröner R, Peters K, von Schnering HG, Nesper R (1998) Z Kristallogr New Cryst Struct 213:671
94. von Schnering HG, Kröner R, Menke H, Peters K, Nesper R (1998) Z Kristallogr New Cryst Struct 213:677
95. Nolas GS, Weakley TJR, Cohn JL (1999) Chem Mater 11:2470
96. von Schnering HG, Carrillo-Cabrera W, Kröner R, Peters EM, Peters K, Nesper R (1998) Z Kristallogr New Cryst Struct 213:679
97. Kröner R, Peters K, von Schnering HG, Nesper R (1998) Z Kristallogr New Cryst Struct 213:667
98. Reny E, Gravereau P, Cros C, Pouchard M (1998) J Mater Chem 8:2839
99. Ramachandran GK, Dong J, Diefenbacher J, Gryko J, Marzke RF, Sanky OF, McMillan PF (1999) J Solid State Chem 145:716
100. Ramachandran GK, McMillan PF, Diefenbacher J, Gryko J, Dong J, Sanky OF (1999) Phys Rev B 60:12294
101. Kawaji H, Horie HO, Yamanaka S (1995) Ishikawa M Phys Rev Lett 74:1427
102. Kawaji H, Iwai K, Yamanaka S, Ishikawa M (1996) Solid State Commun 100:393
103. Ramachandran GK, McMillan PF, Dong J, Sankey OF (2000) J Solid State Chem 154:626
104. Mingos DM (1998) Essential trends in inorganic chemistry. Oxford University Press, Oxford

105. Kröner R, Peters K, von Schnering HG, Nesper R (1998) Z Kristallogr New Cryst Struct 213:669
106. von Schnering HG, Menke H, Kröner R, Peters EM, Peters K, Nesper R Z Kristallogr New Cryst Struct 213:673
107. Kröner R, Peters K, von Schnering HG, Nesper R (1998) Z Kristallogr New Cryst Struct 213:675
108. Kuhl B, Czybulka A, Shuster HU (1995) Z Anorg Allg Chem 621:1
109. Okamoto NL, Tanaka K, Inui H (2006) Acta Mater 54:173
110. Chakoumakos BC, Sales BC, Mandrus DG, Nolas GS (2000) J Alloys Compd 296:80
111. Sales BC, Chakoumakos BC, Jin R, Thompson JR, Mandrus D (2001) Phys Rev B 63:245113
112. Chakoumakos BC, Sales BC, Mandrus DG (2001) J Alloys Compd 322:127
113. Latturner SE, Bryan JD, Blake N, Metiu H, Stucky GD (2002) Inorg Chem 41:3956
114. Latturner SE, Bu X, Blake N, Metiu H, Stucky GJ (2000) J Solid State Chem 151:61
115. Kovnir KA, Zaikina JV, Reshetova LN, Olenev AV, Dikarev EV, Shevelkov AV (2004) Inorg Chem 43:3230
116. Rowe DM (2006) Thermoelectric handbook macro to nano. Taylor & Francis, London
117. Bell LE (2008) Science 321:1457
118. Hudak NS, Amatucci GG (2008) J Appl Phys 103:101301
119. Sootsman JR, Chung DY, Kanatzidis MG (2009) Angew Chem Int Ed 48:8616
120. Dresselhaus MS, Chen G, Tang MY, Yang R, Lee H, Wang D, Ren Z, Fleurial JP, Gogna P (2007) Adv Mater 19:1043
121. Shevelkov AV (2008) Russ Chem Rev 77:1
122. Toberer ES, May AF, Snyder GJ (2010) Chem Mater 22:624
123. Kleinke H (2010) Chem Mater 22:604
124. Christensen M, Abrahamsen AB, Christensen NB, Juranyi F, Andersen NH, Lefmann K, Andreasson J, Bahl CRH, Iversen BB (2008) Nat Mater 7:811
125. Nolas GS, Chakoumakos BC, Mahieu B, Long GJ, Weakley TJR (2000) Chem Mater 12:1947
126. Pacheco V, Bentien A, Carrillo-Cabrera W, Paschen S, Steglich F, Grin Yu (2005) Phys Rev B 71:165205
127. Suekuni K, Takasu Y, Hasegawa T, Ogita N, Udagawa M, Avila MA, Takabatake T (2010) Phys Rev B 81:205207
128. Tanaka T, Onimaru T, Suekuni K, Mano S, Fukuoka H, Yamanaka S, Takabatake T (2010) Phys Rev B 81:165110
129. Zerec I, Keppens V, McGuire MA, Mandrus D, Sales BC, Thalmeier P (2004) Phys Rev Lett 92:185502
130. Bentien A, Pacheco V, Paschen S, Grin Yu, Steglich F (2005) Phys Rev B 71:165206
131. Saramat A, Svensson G, Palmqvist AEC, Stiewe C, Mueller E, Platzek D, Williams SGK, Rowe DM, Bryan JD, Stucky GD (2006) J Appl Phys 99:023708
132. Kuznetsov VL, Kuznetsova LA, Kaliazin AE, Rowe DM (2000) J Appl Phys 87:7871
133. Toberer ES, Christensen M, Iversen BB, Snyder GJ (2008) Phys Rev B 77:075203
134. Martin J, Wang H, Nolas GS (2008) Appl Phys Lett 92:222110
135. Shi X, Yang J, Bai S, Yang J, Wang H, Chi M, Salvador JR, Zhang W, Chen L, Wong-Ng W (2010) Adv Funct Mater 20:755
136. Condron CL, Kauzlarich SM, Gascoin F, Snyder GJ (2006) Chem Mater18:4939
137. Condron CL, Kauzlarich SM, Ikeda T, Snyder GJ, Haarmann F, Jeglič P (2008) Inorg Chem 47:8204
138. Zaikina JV, Mori T, Kovnir KA, Teschner D, Senyshyn A, Schwarz U, Grin Yu, Shevelkov AV (2010) Chem Eur J (in press, DOI: 10.1002/chem.201001990)
139. Gryko J, McMillan PF, Marzke RF, Ramachandran GK, Patton D, Deb SK, Sanke OF (2000) Phys Rev B 62:R7707

140. San-Miguel A, Kéghélian P, Blase X, Mélinon P, Perez A, Itié JP, Polian A, Reny E, Cros C, Pouchard M (1999) Phys Rev Lett 83:5290
141. Moriguchi K, Munetoh S, Shintani A (2000) Phys Rev B 62:7138
142. Smelyansky VI, Tse JS (1997) Chem Phys Lett 264:459
143. Tse JS, Uehara K, Rousseau R, Ker A, Ratcliffe CI, White MA, MacKay G (2000) Phys Rev Lett 85:114
144. Conesa JC, Tablero C, Wahnón P (2004) J Chem Phys 120:6142
145. Demkov AA, Sankey OF, Schmidt KE, Adams GB, O'Keeffe M (1994) Phys Rev B 50:17001
146. Biswas K, Myles CW (2006) Phys Rev B 74:115113
147. Nolas GS, Vanderveer DG, Wilkinson AP, Cohn JL (2002) J Appl Phys 91:8970
148. Rachi T, Tanigaki K, Kumashiro R, Winter J, Kuzmany H (2005) Chem Phys Lett 409:48
149. Beekman M, Sebastian CP, Grin Yu, Nolas GS (2009) J Electron Mater 38:1136
150. Beekman M, Kaduk JA, Gryko J, Wong-Ng W, Shapiro A, Nolas GS (2009) J Alloys Compd 470:365
151. Mott NF (1973) J Solid State Chem 6:348
152. Beekman M, Nolas GS (2006) Physica B 383:111
153. Beekman M, Schnelle W, Borrmann H, Baitinger M, Grin Yu, Nolas GS (2010) Phys Rev Lett 104:018301
154. Latturner SE, Iversen BB, Sepa J, Srdanov V, Stucky G (2001) Phys Rev B 63:125403
155. Gryko J, McMillan PF, Sankey OF (1996) Phys Rev B 54:3037
156. Gryko J, McMillan PF, Marzke RF, Dodokin AP, Demkov AA, Sankey OF (1998) Phys Rev B 57:4172
157. Gryko J, Marzke RF, Lamberton GA Jr, Tritt TM, Beekman M, Nolas GS (2005) Phys Rev B 71:115208
158. Nolas GS, Beekman M, Gryko J, Lamberton GA Jr, Tritt TM (2003) Appl Phys Lett 82:910
159. Li Y, Gao J, Chen N, Liu Y, Luo ZP, Zhang RH, Ma XQ, Cao GH (2008) Physica B 403:1140
160. Madsen GKH, Schwarz K, Blaha P, Singh DJ (2003) Phys Rev B 68:125212
161. Shimizu H, Takeuchi Y, Kume T, Sasaki S, Kishimoto K, Ikeda N, Koyanagi T (2009) J Alloys Compd 487:47
162. Huo D, Sakata T, Sasakawa T, Avila MA, Tsubota M, Iga F, Fukuoka H, Yamanaka S, Aoyagi S, Takabatake T (2005) Phys Rev B 71:075113
163. Avila MA, Huo D, Sakata T, Suekuni K, Takabatake T (2006) J Phys Condens Matter 18:1585
164. Avila MA, Suekuni K, Umeo K, Fukuoka H, Yamanaka S, Takabatake T (2006) Phys Rev B 74:125109
165. Srinath S, Gass J, Rebar DJ, Woods GT, Srikanth H, Nolas GS (2006) J Appl Phys 99:08K902
166. Phan MH, Woods GT, Chaturvedi A, Stefanoski S, Nolas GS, Srikanth H (2008) Appl Phys Lett 93:252505
167. Paschen S, Tran VH, Baenitz M, Carrillo-Cabrera W, Grin Yu, Steglich F (2002) Phys Rev B 65:134435
168. Rachi T, Kitajima M, Kobayashi K, Guo FZ, Nakano T, Ikemoto Y, Kobayashi K, Tanigaki K (2005) J Chem Phys 123:074503
169. Zerec I, Carrillo-Cabrera W, Voevodin V, Sichelschmidt J, Steglich F, Grin Yu, Yaresko A, Kimura S (2005) Phys Rev B 72:045122
170. Zerec I, Yaresko A, Thalmeier P, Grin Yu (2002) Phys Rev B 66:045115
171. Yuan HQ, Grosche FM, Carrillo-Cabrera W, Paschen S, Sparn G, Baenitz M, Grin Yu, Steglich F (2002) J Phys Condens Matter 14:11249
172. Viennois R, Toulemonde P, Paulsen C, San-Miguel A (2005) J Phys Condens Matter 17: L311

173. Rachi T, Tanigaki K, Kumashiro R, Kobayashi K, Yoshino H, Murata K, Fukuoka H, Yamanaka S, Shimotani H, Takenobu T, Iwasa Y, Sasaki T, Kobayashi N, Miyazaki Y, Saito K (2006) J Phys Chem Solids 67:1334
174. Kim JH, Okamoto NL, Kishida K, Tanaka K, Inui H (2006) Acta Mater 54:2057
175. Kim JH, Okamoto NL, Kishida K, Tanaka K, Inui H (2007) J Appl Phys 102:094506
176. Kim JH, Okamoto NL, Kishida K, Tanaka K, Inui H (2007) J Appl Phys 102:034510
177. Fässler TF (1998) Z Anorg Allg Chem 624:569
178. Fässler TF (2003) Chem Soc Rev 32:80
179. Imai M, Nishida K, Kimura T, Yamada K (2002) J Alloys Compd 335:270
180. Yang L, Wang Y, Liu T, Hu TD, Li BX, Ståhl K, Chen SY, Li MY, Shen P, Lu GL, Wang YW, Jiang JZ (2005) J Solid State Chem 178:1773
181. Mudryk Ya, Rogl P, Paul C, Berger S, Bauer E, Hilscher G, Godart C, Noël H (2002) J Phys Condens Matter 14:7991
182. Jaussaud N, Gravereau P, Pechev S, Chevalier B, Ménétrier M, Dordor P, Decourt R, Goglio G, Cros C, Pouchard M (2005) Comptes Rendus Chimie 8:39
183. Alleno E, Maillet G, Rouleau O, Leroy E, Godart C, Carrillo-Cabrera W, Simon P, Grin Yu (2009) Chem Mater 21:1485
184. Melnychenko-Koblyuk N, Grytsiv A, Rogl P, Schmid H, Giester G (2009) J Solid State Chem 182:1754
185. Paschen S, Budnyk S, Köhler U, Prots Yu, Hiebl K, Steglich F, Grin Yu (2006) Physica B 383:89
186. Horie HO, Kikudome T, Teramura K, Yamanaka S (2009) J Solid State Chem 182:129
187. Ramachandran GK, Dong J, Sankey OF, McMillan PF (2000) Phys Rev B 63:033102
188. Fukuoka H, Kiyoto J, Yamanaka S (2003) Inorg Chem 42:2933
189. Imai M, Kikegawa T (2008) Inorg Chem 47:8881
190. Toulemonde P, Adessi Ch, Blase X, San Miguel A, Tholence JL (2005) Phys Rev B 71:094504
191. Toulemonde P, San Miguel A, Merlen A, Viennois R, Le Floch S, Adessi Ch, Blase X, Tholence JL (2006) J Phys Chem Solids 67:1117
192. Wosylus A, Veremchuk I, Schnelle W, Baitinger M, Schwarz U, Grin Yu (2009) Chem Eur J 15:5901
193. Ramachandran GK, McMillan PF, Deb SK, Somayazulu M, Gryko J, Dong J, Sankey OF (2000) J Phys Condens Matter 12:4013
194. Omori M (2000) Mater Sci Eng A 287:183
195. Beekman M, Baitinger M, Borrmann H, Schnelle W, Meier K, Nolas GS, Grin Yu (2009) J Am Chem Soc 131:9642
196. Kishimoto K, Koyanagi T, Akai K, Matsuura M (2007) Jpn J Appl Phys 46:L746
197. Koga K, Suzuki K, Fukamoto M, Anno H, Tanaka T, Yamamoto S (2009) J Electron Mater 38:1427
198. Hayashi M, Kishimoto K, Kishio K, Akai K, Asada H, Koyanagi T (2010) Dalton Trans 39:1113
199. Kanatzidis MG, Pöttgen R, Jeitschko W (2005) Angew Chem Int Ed 44:6996
200. Condron CL, Porter R, Guo T, Kauzlarich SM (2005) Inorg Chem 44:9185
201. Condron CL, Martin J, Nolas GS, Piccoli PMB, Schultz AJ, Kauzlarich SM (2006) Inorg Chem 45:9381
202. Philipp F, Schmidt P (2008) J Cryst Growth 310:5402
203. Böhme B, Aydemir U, Ormeci A, Schnelle W, Baitinger M, Grin Yu (2007) Sci Technol Adv Mater 8:410
204. Böhme B, Guloy A, Tang Z, Schnelle W, Burkhardt U, Baitinger M, Grin Yu (2007) J Am Chem Soc 129:5348

Struct Bond (2011) 139: 143–162
DOI:10.1007/430_2010_20

Published online: 27 July 2010

Hydrogenous Zintl Phases: Interstitial Versus Polyanionic Hydrides

Ulrich Häussermann, Verina F. Kranak, and Kati Puhakainen

Abstract Hydrogen may be incorporated in Zintl phases in two different ways: either hydridic where H is exclusively coordinated by electropositive metals (interstitial hydrides), or as part of the polyanion where it acts as a covalently bonded ligand (polyanionic hydride). Both scenarios provide novel coordination environments and bonding scenarios for the atoms involved. This makes hydrogenous Zintl phases important model systems for fundamental studies of hydrogen–metal interactions. Simultaneously, hydrogen-induced chemical structure and physical property changes provide exciting prospects for materials science.

Keywords: Main group metal hydrides · Polar intermetallic compounds · Polyanionic hydrides · Zintl phases

Contents

U. Häussermann (✉), V.F. Kranak, and K. Puhakainen
Department of Chemistry and Biochemistry, Arizona State University, P.O. Box 871604, Tempe, AZ 85287-1604, USA
e-mail: Ulrich.Haussermann@asu.edu

1 Introduction

Zintl phases consists of an active metal (alkali, alkaline earth, or rare earth; abbreviated as **M** in the following) and a more electronegative p-block metal or semimetal component (Al, Ga, In, Si, Ge, Sn, etc.; abbreviated as **E** in the following). This class of inorganic compounds offers a remarkable wide range of chemical compositions and crystal structures [1–4]. From the beginning of their discovery, Zintl phases intrigued chemists because of their peculiar structural and bonding properties. Lately, they have also attracted attention for their transport (e.g., thermoelectric clathrates or superconducting silicides) and magnetic properties (e.g., magnetocaloric $Gd_5Si_2Ge_2$) [5–7].

Zintl phases bridge structurally and electronically intermetallic, close packed systems with the genuine ionic salts. Similar to ionic salts, an electron transfer from the electropositive component is assumed, leading to reduced p-block atoms which may polymerize to achieve an octet. Thus, the presence of a polyanionic substructure is a characteristic feature of Zintl phases. According to the Zintl–Klemm concept, the atom arrangement of the polyanionic substructure can be rationalized by the new electronic configuration the p-block elements attain after charge transfer. P-block element bonding within the resulting polyanions is considered as covalent – of localized two-center type. Therefore, typical Zintl phases are charge balanced diamagnetic semiconductors, or at least poor metallic conductors. Occasionally, Zintl phases deviate from the Zintl–Klemm concept and display excess or deficiency of electrons with respect to the electronic requirement of the polyanionic substructure. When lowering the electronegativity difference between the active and reduced component, such deviations become more frequent. In addition, multicenter-bonded atom arrangements appear as polyanionic structures, indicating the transition to the fully delocalized metallic systems [8].

Many intermetallic compounds are capable of incorporating hydrogen. In this respect, Zintl phases – bordering true intermetallic systems – are not well investigated. Which role can hydrogen play in connection with Zintl phases? Hydrogen exhibits an electronegativity which is slightly higher than that of p-block elements in Zintl phases, and the following two scenarios are observed: (1) Hydrogen may occur hydridic and exclusively coordinated by active metal ions (in the following those systems will be termed *Zintl phase hydrides* or *interstitial hydrides*). In this case, H acts as an electron sink and reduces the electron count for the p-block component which, according to the Zintl–Klemm concept, should have implications to the polyanionic substructure. (2) Hydrogen may be incorporated in the polymeric anion where it acts as a covalently bonded terminating ligand to a p-block metal atom (those systems will be called *polyanionic hydrides*) [9].

The hydrogen content of both types of hydrides is comparably low; however, chemical structures and physical properties of Zintl phases can be profoundly influenced by the presence of H. In light of the wide range of chemical compositions of Zintl phases, this offers interesting opportunities for H-induced structure and property changes.

2 Synthesis of Hydrogenous Zintl Phases

Hydrogenous Zintl phases can be prepared according to two general schemes (Fig. 1): through reactions of Zintl phases with gaseous hydrogen (hydrogenations, scheme 1) or through reactions of metal hydrides with p-block elements (scheme 2). Hydrogenation reactions may take place at low temperatures (<300°C), while reactions according to scheme 2 typically require high temperatures (>500°C). In the following sections, we explain the two schemes in more detail.

Scheme 1 (Hydrogenations) The formation of interstitial hydrides from hydrogenations can be considered as a first step in a more general sequence of oxidation reactions (Fig. 1a). Hydrogen is incorporated in the M substructure, where it acts hydridic. Accordingly, the polyanion becomes formally oxidized and may reorganize to maintain the octet situation of participating atoms. In Fig. 1a, this is exemplified by a zigzag chain of silicon atoms (as, e.g., in CaSi), which develops into a two-dimensional polyanion when two-bonded "Si^{2-}" (isoelectronic to sulfur) is partially replaced by three-bonded "Si^{1-}" (isoelectronic to phosphorus). As a further step (or instead), a macroscopic segregation of metal hydride and new Zintl phase – now more rich in the p-block component and by this appearing oxidized – may take place. A complete oxidation will correspond to a quantitative decomposition into metal hydride and p-block element.

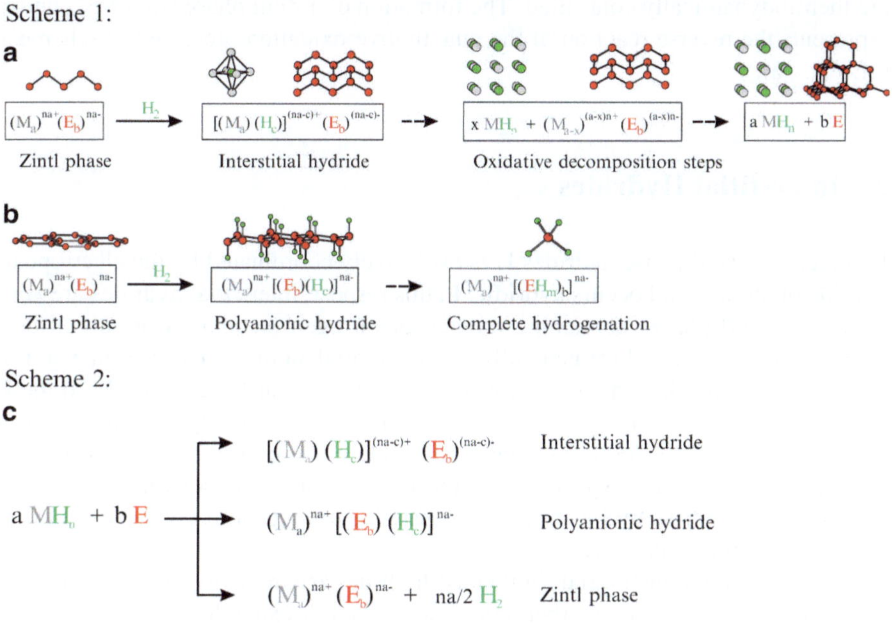

Fig. 1 Formation schemes for hydrogenous Zintl phases. Scheme 1: hydrogenation reactions of Zintl phase precursors leading to (**a**) interstitial and (**b**) polyanionic hydrides. (**c**) Scheme 2: reactions of metal hydrides with p-block elements. For clarity the different kinds of atoms are color coded. The reader is referred to the web version of this article

In polyanionic hydrides, hydrogen is part of the polyanion and thus formally considered as neutral (atomic H). Such hydrides are characterized by the simultaneous occurrence of E–E and E–H bonds. The formation of polyanionic hydrides (Fig. 1b) is conceptually different from interstitial ones. Hydrogen is formally added to, or inserted into, the polyanion of the Zintl phase precursor. Susceptible species for addition would be π-bonded polyanions (e.g., a graphitic layer of "Ga^{1-}" as in $SrGa_2$, cf. Fig. 1b). Insertion implies breaking of precursor E–E bonds and subsequently terminating them by H. In this respect, polyanionic hydrides may be regarded as intermediates toward a complete hydrogenation yielding separated, discrete entities $[EH_n]^{x-}$ that are found in, e.g., borohydrides and alanates.

If and how a Zintl phase will react with hydrogen depends on the thermodynamic (and also kinetic) conditions of a particular system. Low-temperature hydrogenations (<250°C) may proceed in a topotactic way and yield metastable hydrides, whereas high-temperature hydrogenations (>500°C) are more likely to be thermodynamically controlled – which could result in a substantial rearrangement of M and E atoms. Although the scheme presented in Fig. 1a, b suggests a versatile hydrogen-induced polyanion chemistry for Zintl phases, there is little experience and especially no mechanistic studies. Hydrogenous Zintl phases obtained from hydrogenations are attractive because of the inherent possibility to achieve structure and/or property changes reversibly coupled with hydrogen uptake and desorption.

Scheme 2 Typically, reactions of metal hydrides with p-block elements (Fig. 1c) are thermodynamically controlled. The formation of a Zintl phase from MH_n and E represents the reverse reaction of the quantitative oxidation presented in scheme 1 (cf. Fig. 1a).

3 Interstitial Hydrides

In interstitial Zintl phase hydrides H is exclusively coordinated by the electropositive M component and occurs hydridic. In this respect, interstitial hydrides are very similar to Zintl phases containing oxygen or nitrogen (suboxides or subnitrides, respectively) [10–15]. H (or generally, the interstitial atom "Z") acts as an electron sink, which diminishes the number of electrons to be transferred from M to the E component. The M–Z substructure, consisting of Z-centered polyhedra of M atoms that are linked in various ways, may be considered as polycation. The following accounts for important aspects of synthesis, structure, and bonding concerning interstitial Zintl phase hydrides. The discussion is not meant to give an exhaustive overview of representatives.

The formation of interstitial Zintl phase hydrides according to scheme 2 has been especially investigated by Corbett and coworkers. For example, it was shown that many binary M_5Pn_3 phases (M = divalent Ca, Sr, Eu, Yb, Sm; Pn = As, Sb, Bi) with the orthorhombic β-Yb_5Sb_3 structure are actually hydrides $M_5Pn_3H_x$ ($x \approx 1$) [16]. Preparing M_5Pn_3 under H-free conditions affords the hexagonal Mn_5Si_3 structure type [17]. Interestingly, such binary Mn_5Si_3-type systems are able to take up hydrogen. The hydrides either adopt the β-Yb_5Sb_3 structure or remain in the Mn_5Si_3 structure [18]. H attains a quasi-tetrahedral environment by four M atoms in the β-Yb_5Sb_3 structure, whereas coordination in the Mn_5Si_3-type arrangement is octahedral (Fig. 2a).

The application of the Zintl–Klemm concept to M_5Pn_3 phases, $(M^{2+})_5(Pn^{3-})_3 \cdot e^-$, yields one excess electron per formula unit. Therefore, H incorporation in both types of hydrides has been explained as stabilization where the extra electron is trapped in H^- [19].

This is shown in Fig. 3a, b, which depicts the electronic density of states (DOS) for Ca_5Sb_3 and Ca_5Sb_3H, respectively. For Ca_5Sb_3, the Fermi level cuts a cusp in the DOS, whereas the electron precise hydride exhibits a small band gap. H-s states that are dispersed between −5 and −4 eV are substantially involved in the valence band. Note that M_5Pn_3 or M_5Pn_3H do not contain polyanions but fully reduced E

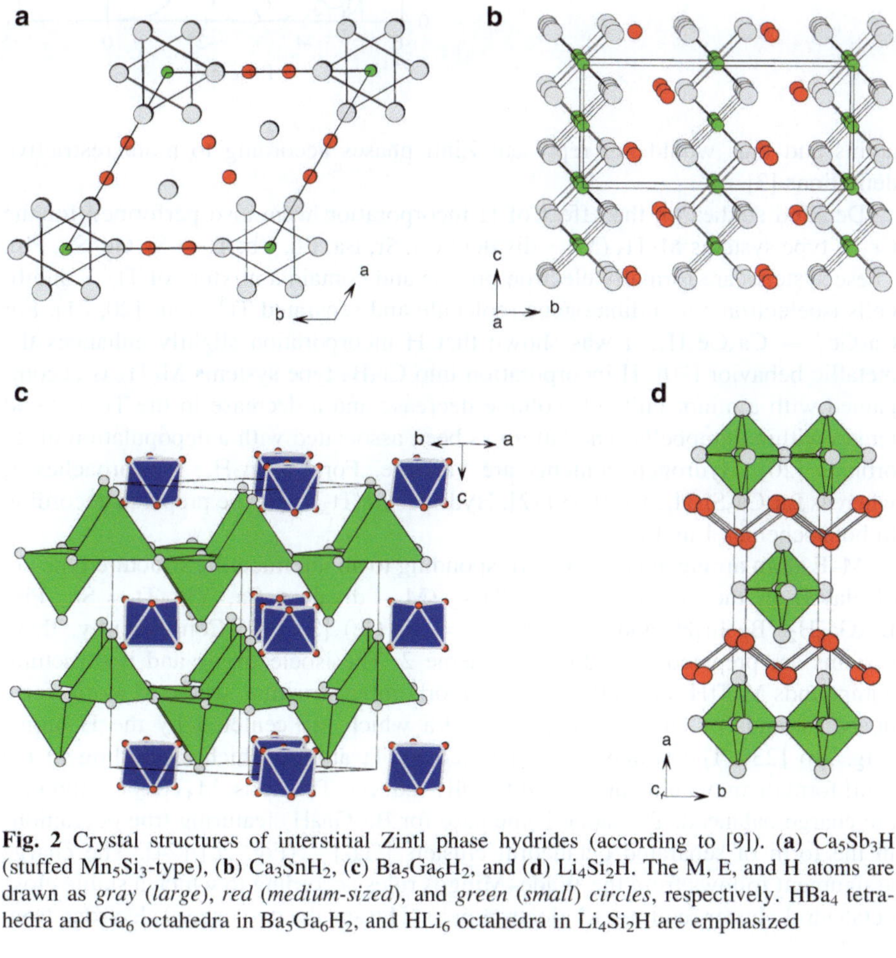

Fig. 2 Crystal structures of interstitial Zintl phase hydrides (according to [9]). (**a**) Ca_5Sb_3H (stuffed Mn_5Si_3-type), (**b**) Ca_3SnH_2, (**c**) $Ba_5Ga_6H_2$, and (**d**) Li_4Si_2H. The M, E, and H atoms are drawn as *gray* (*large*), *red* (*medium-sized*), and *green* (*small*) *circles*, respectively. HBa_4 tetrahedra and Ga_6 octahedra in $Ba_5Ga_6H_2$, and HLi_6 octahedra in Li_4Si_2H are emphasized

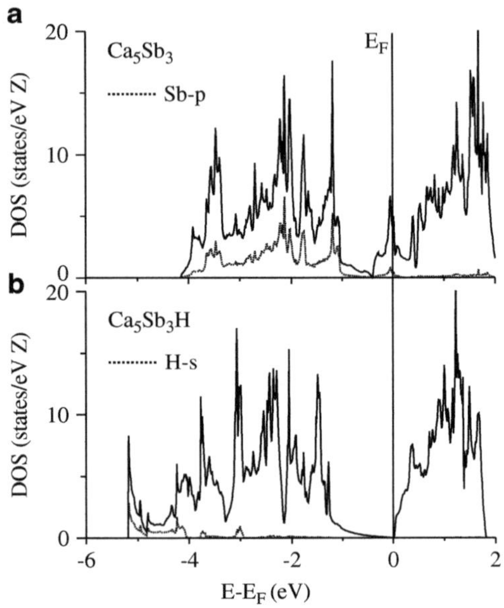

Fig. 3 Electronic density of states (DOS) for Ca_5Sb_3 (**a**) and Ca_5Sb_3H (**b**). The *dotted lines* represent contributions from Sb-p and H-s orbitals, respectively. The Fermi level is marked with a *horizontal line*

atoms and thus would not represent Zintl phases according to more restrictive definitions [3].

Detailed studies on the effect of H incorporation were also performed for the Cr_5B_3 type systems M_5Tt_3 (M = divalent Ca, Sr, Ba, Eu, Yb; Tt = Si, Ge, Sn, Pb). These systems are formally electron precise and contain a mixture of Tt_2^{6-} dumbbells isoelectronic to a dihalogene molecule and separated Tt^{4-} ions [20, 21]. For $Ca_5Ge_3 \leftrightarrow Ca_5Ge_3H_x$, it was shown that H incorporation slightly enhances the metallic behavior [20]. H incorporation into Cr_5B_3 type systems M_5Tt_3 is accompanied with a minor unit cell volume decrease and a decrease in the Tt–Tt bond length within dumbbells. The latter has been associated with a depopulation of π^* orbitals [20]. Hydrogen contents are variable. For $Ca_5Sn_3H_x$, x approaches 1, whereas for $Ca_5Si_3H_x$, x is 0.53 [22]. Hydrides $M_5Tt_3H_x$ can be prepared according to both schemes 1 and 2.

M–E atom arrangements not corresponding to an intermetallic structure type are displayed in the structures of M_3TtH_2 (M = divalent Ca, Yb; Tt = Sn, Pb), $Ba_5Ga_6H_2$, Ba_9In_4H, and Li_4Tt_2H (Tt = Si, Ge) [23–26]. Consequently, these systems are prepared according to scheme 2. The isoelectronic and isostructural compounds M_3TtH_2 crystallize in an orthorhombic structure, where M atoms form double strands of edge sharing tetrahedra which are centered by the H atoms (Fig. 2b) [23]. The strands are separated by Tt atoms, which, according to the Zintl formalism, would correspond to fully reduced Tt^{4-} ions. M_3TtH_2 compounds are charge balanced. The same is the case for $Ba_5Ga_6H_2$ featuring true polyanions in the form of separated octahedral clusters $[Ga_6]^{8-}$ (Fig. 2c) [24]. The charge assignment follows from the Wade–Mingos rules according to which a stable close octahedral cluster requires 13 electron pairs. H traps the two excess electrons from the M component and provides charge balance, $(Ba^{2+})_5[Ga_6]^{8-}(H^-)_2$, in agreement with the Zintl–Klemm concept. The arrangement of Ba atoms provides tetrahedral voids which are centered by H atoms. HBa_4 tetrahedra are corner-connected and organized in layers. In the recently discovered hydride, Ba_9In_4H hydrogen atoms are coordinated by six Ba atoms and HBa_6 octahedra are connected by opposite corners to form chains [25]. Indium occurs isolated and, thus, the Zintl–Klemm concept yields an electron-deficient compound $(Ba^{2+})_9(In^{5-})_4(H^-)$. This is unusual because the presence of H enhances electron deficiency instead of reducing or even removing it.

The orthorhombic crystal structure of Li_4Tt_2H consists of zigzag chains formed by the Tt atoms which are sandwiched between layers of H-centered corner sharing Li_6 octahedra (Fig. 2d) [26]. The charge assignment according to the Zintl–Klemm concept would be $(Li^+)_4[Tt_2]^{3-}H^-$. The electron requirement for a zigzag chain formed by main group elements can be considered flexibly. A combination of single bonds and fully occupied lone pairs requires six electrons per atom. For Tt atoms, this would imply a charge of −2. Alternatively, a charge of −1 results when assuming that the Tt-p orbital perpendicular to the chain is singly occupied and thus the chain is additionally π bonded. Electron counts between 5 and 6 (or charges between −1 and −2) for chain forming Tt atoms could then be accounted for by a partially filled antibonding π* band. This simple analysis predicts Li_4Tt_2H to be a metallic conductor, which indeed has been verified in a recent electronic structure analysis [27].

A zigzag chain formed by Si atoms is also at the heart of the system CaSi ↔ $CaSiH_{1+x}$, which is a beautiful example of hydrogen-induced polyanion chemistry in Zintl phases (Fig. 4) [28, 29]. CaSi crystallized in the orthorhombic CrB type (Fig. 4a). An electron count of six per Si atom (Si^{2-}) suggests singly bonded Si chains. The Si–Si distance in CaSi is 2.47 Å, which is considerably larger than in Li_4Si_2H (2.39 Å). However, it is difficult to associate distances with actual bonding situations in Zintl phases (here a single-bonded chain vs. a partially double-bonded

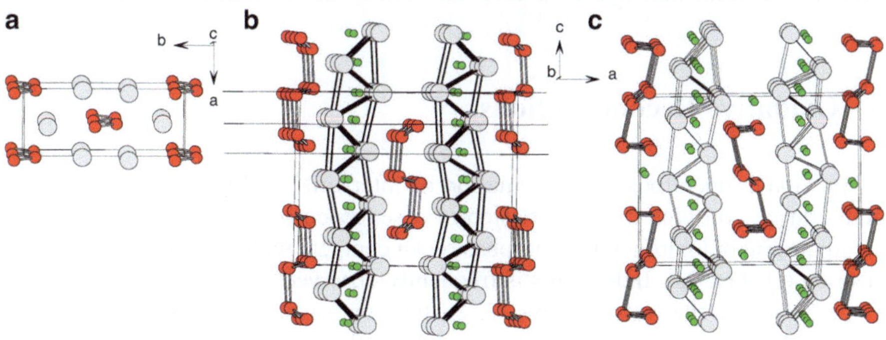

Fig. 4 The crystal structures of CaSi (**a**), $CaSiH_{\sim 1.0}$ (**b**), and $CaSiH_{\sim 1.2}$ (**c**) according to [9]. The Ca, Si, and H atoms are drawn as *gray* (*large*), *red* (*medium-sized*), and *green* (*small*) *circles*, respectively

one). This is because the formal charge on the E atom and the counteraction environment also influence bond lengths in polyanionic arrangements.

Depending on reaction conditions, two distinct hydrides, $CaSiH_1$ and $CaSiH_{1.2}$, can be obtained [30]. In these hydrides, Ca atoms are slightly shifted to form a two-dimensional array of edge sharing tetrahedra, which are centered by H atoms. The Si chains in turn appear mutually rotated, which gives new interchain Si–Si contacts, linking entities of three chains to new polyanionic units (Fig. 4b). $CaSiH_{1.2}$ contains an additional, not fully occupied, H position (Fig. 4c) [30]. This position is located outside the tetrahedral array of Ca atoms and close to one type of Si atoms (Si–H = 1.82 Å). Both hydrides can be completely dehydrogenated to CaSi.

H incorporation in CaSi induces the formation of new Si–Si bonds; however, the application of the Zintl–Klemm formalism to CaSiH is problematic. The presence of H^- would reduce the electron count to an average of 5 per Si atom. This can be compensated for by either introducing π-bonding between two-connected Si atoms within chains or the creation of three-bonded moieties. The new polyanions consist of a mixture of two- and three-bonded Si atoms (ratio 2:1). A closer look reveals some peculiarities. The Si–Si distance within zigzag chain units (2.23 and 2.34 Å, two types of chains) is shortened considerably with respect to the precursor CaSi (2.47 Å), while chain-linking distances are rather long (2.64 Å). As a matter of fact, the intrachain Si–Si distances in CaSiH are unusually short for Si polyanions in Zintl phases. It appears that besides interchain linkage, H incorporation also introduces intrachain π-bonding.

Interestingly, when going from CaSiH to $CaSiH_{1.2}$, interatomic distances within the Si polyanion widen. They become 2.32 and 2.58 Å within zigzag chains and 2.90 Å for the linking distance. This is not expected if the additional H atoms acted interstitially and raises doubts into their hydridic nature. In summary, the system $CaSiH_{1+x}$ displays a highly interesting structural chemistry and possibly bonding properties that need further clarification. Further excitement about hydrogenous CrB-type silicides comes from recent reports that SrSi, BaSi, and EuSi incorporate substantially larger amounts of H than CaSi [31]. In contrast to $CaSiH_{1+x}$, the structures of these new hydrides have not been completely characterized.

4 Oxidative Decompositions

As previously mentioned, the formation of interstitial hydrides from hydrogenations can be considered as a first step of a sequence of oxidation reactions. The incorporation of hydrogen in Zintl phases can compete with their oxidative decomposition (cf. Fig. 2a). Indeed this is frequently observed.

Ca_2Si adopts the orthorhombic Co_2Si type with isolated, formally completely reduced, Si atoms (Si^{4-}). Depending on the reaction conditions, the hydrogenation of Ca_2Si leads to three different products: (1) a mixture of crystalline CaH_2 and $Ca_5Si_3H_x$, (2) a mixture of CaH_2 and non-hydrogenous CaSi, or (3) amorphous Ca_2SiH_x ($x \approx 2.41$). The amorphous product requires the application of high hydrogen pressures (above 50 bar), and Ca_2Si reforms after high-temperature dehydrogenation. The H-induced amorphization of Ca_2Si has been explained by a "chemical frustration." Interdiffusion of Ca and Si is too slow toward a macroscopic segregation into CaH_2 and silicide phase (or elemental Si), and thus a metastable, intermediate phase is formed [32]. Also Ca_5Si_3 (or rather $Ca_5Si_3H_x$) decomposes upon hydrogenation at pressures above 30 bar [22] (note that the formation of $Ca_5Si_3H_{0.53}$ already took place at 1 bar hydrogen pressure).

During the course of exploring polyanionic hydrides, we frequently observed oxidative decompositions of Zintl phases containing Al or Ga polyanions. For example, both Ca_3Ga_5 and Li_3Al_2 have imbalanced (electron deficient) Ga and Al polyanions. They were expected to readily react with hydrogen to yield electron precise Ca_3Ga_5H ($[Ga_5H]^{6-}$) and Li_3Al_2H ($[Al_2H]^{3-}$), respectively. However, instead of polyanonic hydride formation, hydrogenation leads to the decompositions $CaGa_4 + CaH_2$ and LiAl + LiH. An interesting feature is the low temperatures (below 200°C) at which these reactions occur. The "record holder" is $SrZn_2$,

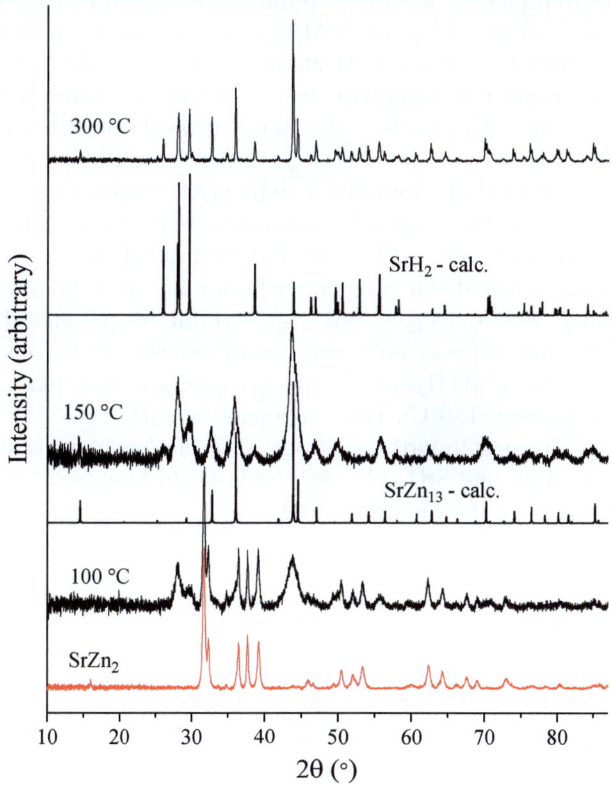

Fig. 5 Oxidative decomposition of $SrZn_2$ ($CeCu_2$ type) into SrH_2 and $SrZn_{13}$ ($NaZn_{13}$ type) upon hydrogenation. X-ray powder patterns (Cu-Kα) of the product mixture are shown for synthesis temperatures of 100, 150, and 300°C (reaction time 24 h, pressure 50 bar)

which in a hydrogen atmosphere transforms at 75°C over a period of 2 weeks quantitatively into a nanocrystalline $SrZn_{13}$ and SrH_2. Higher temperatures (300°C) afford the mixture highly crystalline within several hours (Fig. 5).

5 Polyanionic Hydrides

In contrast to Zintl phase hydrides that contain H interstitially and hydridic in a M atom environment, polyanionic hydrides are compounds in which hydrogen atoms are part of an extended (polymeric) anion. Thus, polyanionic hydrides are distinguished by the simultaneous presence of E–E and E–H bonds.

The first reported representative of a polyanionic hydride is trigonal $SrAl_2H_2$, which was obtained from hydrogenating the precursor phase $SrAl_2$ [33]. $SrAl_2$ ($CeCu_2$ type) features a three-dimensional polyanionic network in which each Al atom is surrounded by four neighbors in a distorted tetrahedral fashion. This structure is electron precise according to the Zintl–Klemm concept (four-bonded Al^- counterbalanced by Sr^{2+}). In $SrAl_2H_2$, the dimension of the polyanionic framework is reduced to two: the Al atoms are arranged as puckered graphitic layers with three nearest Al neighbors, while the vacant coordination is taken by hydrogen. Isostructural MGa_2H_2 (M = Sr, Ba) is obtained from differently structured precursors [34].

The formation of $SrAl_2H_2$ from $CeCu_2$-type $SrAl_2$ follows an insertion scheme where bonds are cut and subsequently terminated by H, whereas the formation of MGa_2H_2 from AlB_2-type MGa_2 (M = Sr, Ba) corresponds to the addition scheme (Fig. 6). Hydrogen is added to a three-bonded, formally sp^2 hybridized and unsaturated Ga^{1-} entity, which changes into a sp^3 hybridized one upon binding to H. Importantly, addition imposes only very small changes to the structure of the precursor Zintl phase [34]. Hydrogenation reactions can take place at mild, low temperature conditions (~150°C). The compounds $SrAl_2H_2$, $SrGa_2H_2$, and $BaGa_2H_2$ are not stable with respect to further hydrogenation and/or decomposition ($SrAl_2H_2$ to the alanate Sr_2AlH_7 and SrH_2 [35, 36]; MGa_2H_2 to MH_2 and MGa_4 [34]), and they cannot be prepared from mixtures MH_2 and E.

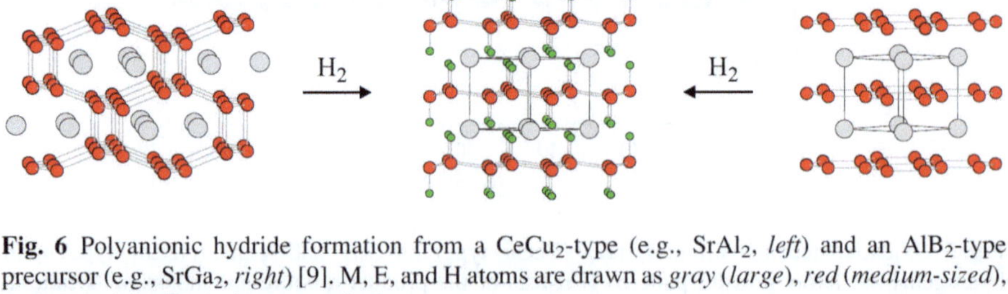

Fig. 6 Polyanionic hydride formation from a $CeCu_2$-type (e.g., $SrAl_2$, *left*) and an AlB_2-type precursor (e.g., $SrGa_2$, *right*) [9]. M, E, and H atoms are drawn as *gray* (*large*), *red* (*medium-sized*), and *green* (*small*) *circles*, respectively

The two-dimensional polyanion $[Tr_2H_2]^{2-}$ (Tr = Al, Ga) is formally electron precise and isoelectronic to polysilane/polygermane. Al-H and Ga-H distances are very similar and approximately 1.70 Å. The electronic structure and dynamic properties of MTr_2H_2 systems have been analyzed in detail by inelastic neutron scattering (INS) spectroscopy and first principles calculations [37, 38]. Hydrides MTr_2H_2 are not semiconductors but have their Fermi level located at a pronounced pseudogap in the DOS (Fig. 7) [34]. H-s states strongly hybridize with Tr-s and Tr-p bands, which indicates that H is (polarized) covalently bonded to the Tr framework and thus part of the two-dimensional polyanion $[Tr_2H_2]^{2-}$.

From a chemical point of view, it is easy to identify Zintl phases with polyanions that should be susceptible to H addition. Such polyanions will be either formally unsaturated (i.e., displaying a π-bonding contribution) or be electronically "imbalanced." In this respect, the ubiquitous ternary compounds MTrTt (M = Ca, Sr, Ba; Tr = Al, Ga, In; Tt = Si, Ge, Sn) have especially proven to be fruitful (Fig. 8). Fifteen representatives are known today [40, 41]. Their structures derive from the simple AlB_2 type where Tr and Tt atoms are commonly distributed over the site of the boron atoms forming planar hexagon layers, and M atoms occupy the "intercalating" position. Therefore, when considering compounds MTrTt as Zintl phases, they are charge imbalanced (the optimum electron count for a polyanionic graphitic layer is eight electrons per formula unit) and are metals. The antibonding π* band of the polyanionic graphitic layer is partially occupied. Due to the different energies of Tr and Tt-p_z orbitals, it has typically a larger Tr-p_z contribution [42].

The π* band in MTrTt is associated with a number of peculiarities. It is responsible for a soft mode, the Tr–Tt out-of-plane mode, which expresses an

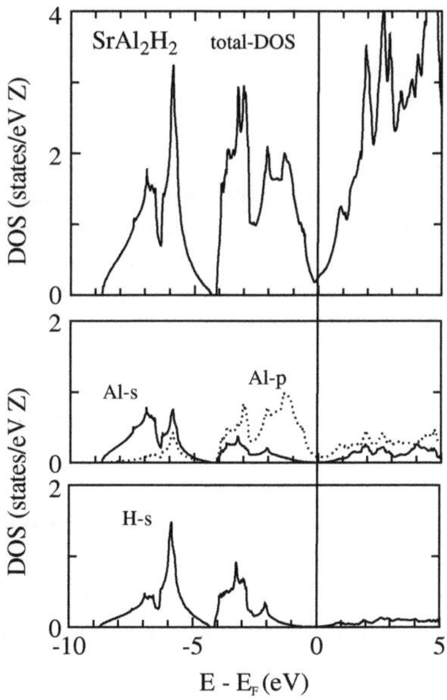

Fig. 7 Electronic density of states (DOS) for $SrAl_2H_2$ [34]. *Top*: total DOS; *bottom*: Al-s, Al-p, and H-s site projected DOS. The Fermi level is marked by the *horizontal line*

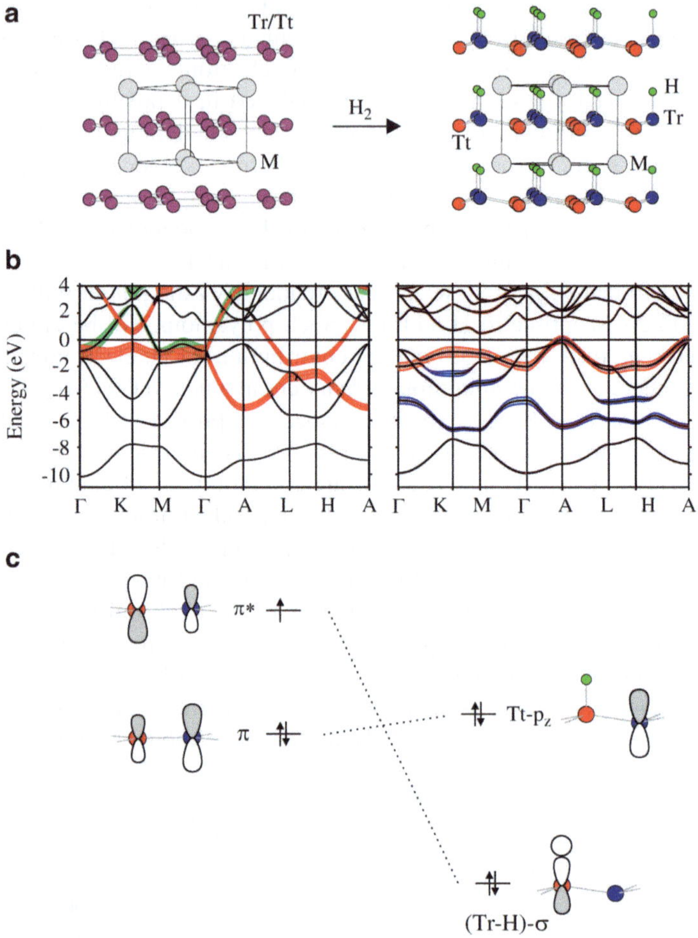

Fig. 8 (**a**) Polyanionic hydride formation from AlB_2 nine-electron type precursors MTrTt (M atoms *gray*, Tr/Tt disordered site *purple*, Tr atoms *blue*, Tt atoms *red*, H atoms *green*). (**b**) Band structure for CaAlSi (*left*, *red* "fat bands" indicate the π and π* bands, the *green* fat band corresponds to a Ca-d band), and CaAlSiH (*right*, the *red* fat band indicates the Al-p_z band, the *blue* fat band is the Al-H band) exemplifying the electronic structure change (metal-semiconductor transition) upon hydrogenation of MTrTt [39]. (**c**) Sketch of the electronic structure change induced by the hydrogenation of MTrTt [39] (for interpretation of the references to color in this figure legend, the reader is referred to the web version of this article)

inherent structural instability toward layer puckering [43, 44]. Indeed, many systems MTrTt display commensurately and incommensurately modulated structures where hexagon layers are slightly corrugated [41]. The Tr–Tt soft mode is further an important ingredient to the superconducting properties of MTrTt systems. This has been first realized for the silicide (Tt = Si) systems [6, 45, 46] but later extended to germanides and even stannides [41]. The partially filled π* band is also connected to the capability of MTrTt systems to incorporate H and form hydrides MTrTtH (Fig. 8a). Hydrogenation (i.e., scheme 1) induces a metal(superconductor)–nonmetal transition by removing this band and turning it into an energetically low lying Tr–H bonding band [39]. In turn, the fully occupied bonding π band in MTrTt changes to a weakly dispersed band with Tt-p_z (lone pair) character in the hydrides, which becomes located below the Fermi level (Fig. 8b, c) [39]. Monohydrides MTrTrH are all electron precise semiconductors. Depending on composition, band gaps vary between 0.1 and 0.8 eV [47–49].

The structure of MTrTtH (SrAlSiH type) is closely related to the $SrAl_2H_2$ type. It displays puckered hexagon nets in which [Tr–H] entities and isoelectronic Tt atoms are arranged strictly alternating. According to the Zintl–Klemm concept, the monohydrides MTtTrH contain layered polyanions $[TrTtH]^{2-}$ with three-bonded $[Tr–H]^{1-}$ and three-bonded, lone pair carrying, Tt^{1-} entities. The charge of the polyanion is balanced by M^{2+} cations. In the SrAlSiH type, the center of inversion present in $SrAl_2H_2$ is lost and space group symmetry is reduced to $P3m1$ [47]. The most conspicuous difference between the two structures is that in MTrTtH, M is only coordinated by three H atoms. In addition, [Tr–H] entities are well separated, whereas in the $SrAl_2H_2$ type H atoms approach each other at a distance of below 3 Å. It is important to note that for Tr/Tt disordered precursors, hydride formation is accompanied by a rearrangement of Tr and Tt atoms within the hexagon layers. Typically, MTrTt hydrogenations require temperatures above 275°C for appreciable kinetics.

The great compositional flexibility of polyanionic hydrides MTrTtH allows for comparative studies. For example, Fig. 9 compiles calculated (zero-temperature) formation energies for the reaction MTrTt + ½ H_2 → MTrTtH considering all compositional combinations leading to monohydrides and assuming MTrTt with

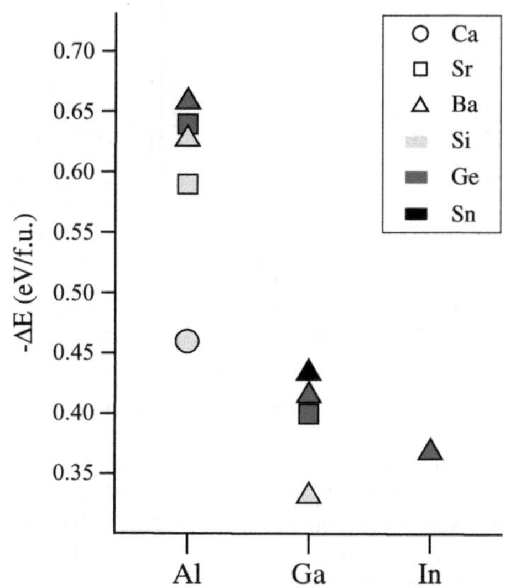

Fig. 9 Computed zero-temperature formation energy for the reaction MTrTt + $1/2H_2$ = MTtTrH according to [49] [MTrTt is assumed with an ordered AlB_2 (i.e., SrPtSb) structure]

an ordered AlB_2 (i.e., SrPtSb) structure [49]. The stability of MTrTtH increases with increasing size of M, or equivalently with increasing separation of polyanionic layers. Furthermore, hydrides with Tr = Al are distinguished by their high formation energy. They afford the only Ca compound (CaAlSiH) and display a substantial higher thermal stability compared to Tr = Ga, In compounds (desorption temperatures can be above 600°C for Tr = Al, while they are typically below 400°C for Tr = Ga, In). Weakly stable BaInGeH is a rare example of a solid-state compound containing the elusive In–H bond [50].

Furthermore, monohydrides MTrTtH represent interesting model systems for fundamental studies of Tr–H interactions by vibrational spectroscopy with INS. This is because of the unique way by which separated entities [Tr–H] are embedded in a flexible environment of M and Tt [37, 50, 51]. Spectra of MTrTtH are characterized by a weakly dispersed Tr–H bending and stretching modes, and well-resolved overtones. The latter allow the study of anharmonicities. The fundamental stretching modes are a direct measure of the Tr–H bond strength. The degenerate bending modes – where H displaces toward M atoms – reflect indirectly the strength of M–H interactions and will stiffen (i.e., shift to higher frequencies) for stronger M–H interactions. As an example, the INS spectrum of SrAlSiH (Fig. 10) displays the Al-H bending and stretching modes at approximately 850 and 1,200 cm^{-1}, respectively. Noteworthy is that [Tr–H] stretching frequencies for polyanionic hydrides are exceptionally low compared to "traditional" systems displaying terminal Tr–H bonds (e.g., molecular species TrH_n ($n = 1, 2, 3$) [52–55] or tetrahedral complexes TrH_4^- occurring in, e.g., $NaAlH_4$ and $NaGaH_4$ [56, 57]).

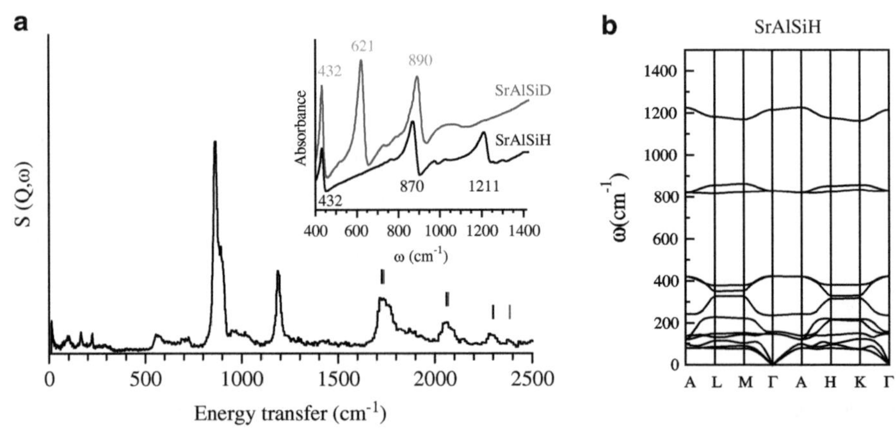

Fig. 10 (**a**) Inelastic neutron scattering (INS) spectrum of SrAlSiH, with the IR spectrum of SrAlSiH and SrAlSiD *inset*. The *vertical bars* in the INS spectrum denote the overtones. The difference between the *black* and *gray bar* at approximately 2,300 and 2,400 cm^{-1} expresses anharmonicity of the Al-H stretchning mode. (**b**) Calculated phonon dispersion curves for SrAlSiH. Figures are taken from [37]

The rather weak terminal Tr–H bond in polyanionic hydrides appears surprising considering their thermal stability. A recent bonding analysis for MAlSiH (M = Ca, Sr, Ba) revealed that H actually occurs hydridic (as in interstitial hydrides) giving rise to strong M–H interactions [39]. The Tr–H bond is probably best described as dative bond to nominally zero-charged Tr, i.e., [Tr ← H^{1-}] instead of the [$Tr^{1-}\cdot$ ·H]. The latter being the formal border situation according to the Zintl–Klemm concept.

In general, polyanionic hydrides offer new chemical environments for E atoms where they are coordinated simultaneously to additional E and H atoms. Hitherto, representatives of polyanionic hydrides are limited to examples where hydrogen is attached to one of the group 13 metals Al, Ga, or In. It is not yet clear whether polyanions with E–H bonds can also be obtained for more electronegative elements E, such as Si. In this respect, $CaSiH_{1.2}$ (Fig. 4c) may represent an interesting bridge [30]. In principle, the Zintl–Klemm concept allows for the identification of

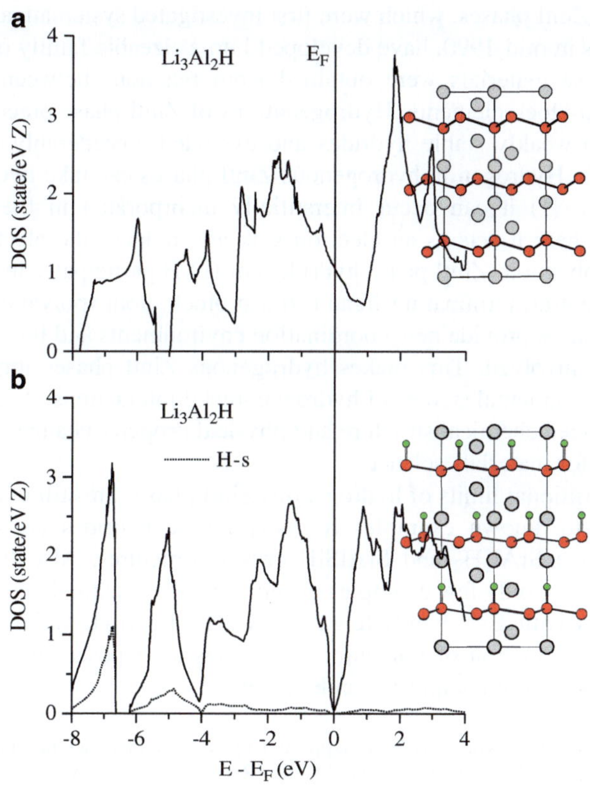

Fig. 11 Electronic density of states (DOS) for Li_3Al_2 (**a**) and hypothetical Li_3Al_2H (**b**). The Fermi level is marked by the *horizontal line*. The corresponding crystal structures are shown on the *right-hand part* of the figure. Li, Al, and H atoms are drawn as *gray* (*large*), *red* (*medium-sized*), and *green* (*small*) *circles*, respectively

unsaturated or electron deficient precursor polyanions which are susceptible for H addition. For example, in rhombohedral Li_3Al_2 polyanions correspond to puckered hexagon layers with an electron requirement of five per atom (i.e., formally Al^{2-}). This is not achieved in Li_3Al_2. It is clearly seen in its DOS that the phase is electron deficient by one electron per formula unit (Fig. 11a). Now, one could envision an electron precise monohydride Li_3Al_2H with a polyanion composed of three-bonded $[Al]^{2-}$ and $[Al\text{-}H]^{1-}$ moieties, which is isostructural and isoelectronic to that in the SrAlSiH structure (Fig. 11b). However, the formation of hydrides is not easily predictable. As discussed earlier, hydrogenation of Li_3Al_2 leads to the oxidative decomposition of LiAl + LiH.

6 Conclusions

Hydrogenous Zintl phases, which were first investigated systematically by Corbett and coworkers in mid-1990, have developed into a sizeable family of compounds. Originally these materials were obtained from reactions between active metal hydrides and p-block elements. Hydrogenations of Zintl phase precursors enabled then access to weakly stable hydrides and extended considerably the variety of representatives. Hydrogen in hydrogenous Zintl phases can take two conceptually different roles: (1) it can occur interstitially incorporated in the active metal substructure where it acts as an electron sink and reduces the electron count for the p-block polyanion (Zintl phase hydrides) or (2) H participates in the polyanion as formally neutral, terminating ligand to a p-block atom (polyanionic hydrides) [9]. Both scenarios provide new coordination environments and bonding situations for the atoms involved. This makes hydrogenous Zintl phases important model systems for fundamental studies of hydrogen–metal interactions. At the same time, hydrogen-induced chemical structure and physical property changes provide exciting prospects for materials science.

The compositional limits of hydrogenous Zintl phases are still essentially unexplored. Hitherto, known examples for polyanionic hydrides are compositional variations of the $SrAl_2H_2$ and SrAlSiH crystal structures. Also for Zintl phase hydrides, there are only few examples of representatives actually containing genuine polyanions with E–E bonds (e.g., $Ba_5Ga_6H_2$, Li_4Si_2H, CaSiH). The factors leading to the formation of polyanionic hydrides as opposed to interstitial Zintl phase hydrides are not completely understood.

Acknowledgments This work has been supported by the National Science Foundation grants DMR-0638826 and CHE-0742006, and the Swedish Research Council (VR). It has also made use of the Manuel Lujan, Jr. Neutron Scattering Center at Los Alamos National Laboratory, which is funded by the Department of Energy's Office of Basic Energy Sciences. Los Alamos National Laboratory is operated by Los Alamos National Security, LLC, under DOE Contract DE-AC52-06NA25396.

References

1. Kauzlarich SM (1996) Chemistry, structure and bonding of Zintl phases and ions. VCH, Weinheim
2. Schäfer H, Eisenmann B, Müller W (1973) Zintl phases – transitions between metallic and ionic bonding. Angew Chem Int Ed 12:694–712
3. Nesper R (1990) Structure and chemical bonding in Zintl-phases containing lithium. Prog Solid State Chem 20:1–45
4. Sevov SC (2002) Zintl Phases. In: Westbrook JH, Fleischer RL (eds) Intermetallic compounds – principles and practice, vol 3. Wiley, Chichester
5. Nolas GS, Cohn JL, Slack GA, Schujman SB (1998) Semiconducting Ge clathrates: promising candidates for thermoelectric applications. Appl Phys Lett 73:178–180
6. Imai M, Nishida K, Kimura T, Abe H (2002) Superconductivity of $Ca(Al_{0.5}Si_{0.5})_2$, a ternary silicide with the AlB_2 structure. Appl Phys Lett 80:1019–1021
7. Pecharsky VK, Gschneidner KA (1997) Giant magnetocaloric effect in $Gd_5(Si_2Ge_2)$. Phys Rev Lett 78:4494–4497
8. Corbett JD (2000) Polyanionic clusters and networks of the early p-element metals in the solid state: beyond the Zintl boundary. Angew Chem Int Ed 39:670–690
9. Häussermann U (2008) Coexistence of hydrogen and polyanions in multinary main group element hydrides. Z Kristallogr 223:628–635
10. Wang LM, Tang ZJ, Lorenz B, Guloy AM (2008) Remarkable rare-earth metal silicides oxides with planar Si_6 rings. J Am Chem Soc 130:11258–11259
11. Guloy AM, Corbett JD (1994) Exploration of the intersitital derivatives of La_5Pb_3 (Mn_5Si_3-type). J Solid State Chem 109:352–358
12. Xia SQ, Bobev S (2007) On the existence of Ca_2Bi-crystal and electronic structure of Ca_4Bi_2O. J Alloys Compd 427:67–72
13. Lulei M (1997) $Ba_{11}KX_7O_2$ (X = P, As): two novel Zintl phases with infinite chains of oxygen centered Ba_6 octahedra, isolated X^{3-} and X_2^{4-} anions. Z Allg Anorg Chem 623:1796–1802
14. Wendorff M, Röhr C (2006) $Ba_{11}In_6O_3$: an indide oxide with novel $[In_6]$ building units. Z Anorg Allg Chem 632:1792–1798
15. Boss M, Petri D, Pickhard F, Zönnchen P, Röhr C (2005) New barium antimonide oxides containing Zintl ions $[Sb]^{3-}$, $[Sb_2]^{4-}$, and $[\infty^1]$ $[Sb_n]^{n-}$ Z. Anorg Allg Chem 631:1181–1190
16. Leon-Escamilla EA, Corbett JD (1994) Compounds of alkaline-earth and divalent rare-earth metals stabilized by hydrogen impurities – the Yb_5Sb_3 and Mn_5Si_3 structure types for pnictides. J Alloys Compd 206:L15–L17
17. Leon-Escamilla EA, Corbett JD (1998) Hydrogen stabilization. Nine isotypic orthorhombic A_5Pn_3H phases (among A = Ca, Sr, Ba, Sm, Eu, Yb; Pn = Sb, Bi) formerly described as binary beta-Yb_5Sb_3-type compounds. J Alloys Compd 265:104–114
18. Leon-Escamilla EA, Corbett JD (2006) Hydrogen in polar intermetallics. Binary pnictides of divalent metals with Mn_5Si_3-type structures and their isotypic ternary hydride solutions. Chem Mater 18:4782–4792
19. Corbett JD, Leon-Escamilla EA (2003) Role of hydrogen in stabilizing new hydride phases or altering old ones. J Alloys Compd 356:59–64
20. Leon-Escamilla EA, Corbett JD (2001) Hydrogen impurity effects. A_5Tt_3 intermetallic compounds between A = Ca, Sr, Ba, Eu and Tt = Si, Ge, Sn with Cr_5B_3-like structures that are stable both as binary and as ternary hydride and fluoride phases. J Solid State Chem 159:149–162
21. Leon-Escamilla EA, Corbett JD (2001) Hydrogen impurity effects. A_5Tt_3Z intermetallic compounds between A = Ca, Sr, Ba, Eu, Yb and Tt = Sn, Pb with Cr_5B_3-like structures that are stabilized by hydride or fluoride Z. Inorg Chem 40:1226–1233
22. Wu H, Zhou W, Udovic TJ, Rush JJ, Yilderim T (2008) Structural variations and hydrogen storage properties of Ca_5Si_3 with Cr_5B_3-type structure. Chem Phys Lett 460:432–437

23. Huang BQ, Corbett JD (1997) Intermetallic hydrides as Zintl phases: A_3TtH_2 compounds (A = Ca, Yb; Tt = Sn, Ph) and their structural relationship to the corresponding oxides. Inorg Chem 36:3730–3734
24. Henning RW, Leon-Escamilla EA, Zhao JT, Corbett JD (1997) Stabilization by hydrogen. Synthetic and structural studies of the Zintl phase $Ba_5Ga_6H_2$. Inorg Chem 36:1282–1285
25. Wendorff M, Scherer H, Röhr C (2010) The indide-hydride Ba_9In_4H: synthesis, crystal structure, NMR spectroscopy, chemical bonding. Z Anorg Allg Chem 636:1038–1044
26. Wu H, Hartman MR, Udovic TJ, Rush JJ, Zhou W, Bowman RC, Vajo JJ (2007) Structure of the novel ternary hydrides Li_4Tt_2D (Tt = Si and Ge). Acta Crystallogr B 63:63–68
27. Wu H, Zhou W, Udovic TJ, Rush JJ, Yildirim T, Hartman MR, Bowman RC, Vajo JJ (2007) Neutron vibrational spectroscopy and first-principles calculations of the ternary hydrides $Li_4Si_2H(D)$ and $Li_4Ge_2H(D)$: electronic structure and lattice dynamics. Phys Rev B 76:224301
28. Aoki M, Ohba N, Noritake T, Towata S (2004) Reversible hydriding and dehydriding properties of CaSi: potential of metal silicides for hydrogen storage. Appl Phys Lett 85:387–388
29. Ohba N, Aoki M, Noritake T, Miwa K, Towata S (2005) First-principles study of a hydrogen storage material CaSi. Phys Rev B 72:075104
30. Wu H, Zhou W, Udovic TJ, Rush JJ, Yildirim T (2006) Structure and hydrogen bonding in $CaSiD_{1+x}$: issues about covalent bonding. Phys Rev B 74:224101
31. Armbruster M, Wörle M, Krumeich F, Nesper R (2009) Structure and properties of hydrogenated Ca, Sr, Ba, and Eu silicides. Z Allg Anorg Chem 635:1758–1766
32. Wu H, Zhou W, Udovic TJ, Rush JJ (2007) Hydrogen storage in a novel destabilized hydride system, Ca_2SiH_x: Effects of amorphization. Chem Mater 19:329–334
33. Gingl F, Vogt T, Akiba E (2000) Trigonal $SrAl_2H_2$: the first Zintl phase hydride. J Alloys Compd 306:127–132
34. Björling T, Noreus D, Häussermann U (2006) Polyanionic hydrides from polar intermetallics AeE_2 (Ae = Ca, Sr, Ba; E = Al, Ga, In). J Am Chem Soc 128:817–824
35. Zhang QA, Nakamura Y, Oikawa K, Kamiyama T, Akiba E (2002) Synthesis and crystal structure of Sr_2AlH_7: a new structural type of alkaline earth aluminum hydride. Inorg Chem 41:6547–6549
36. Zhang QA, Nakamura Y, Oikawa K, Kamiyama T, Akiba E (2003) Addition and correction. Inorg Chem 42:3152
37. Lee MH, Sankey OF, Björling T, Moser D, Noréus D, Parker SF, Häussermann U (2007) Vibrational properties of the polyanionic hydrides $SrAl_2H_2$ and SrAlSiH: new insights into Al-H interactions. Inorg Chem 46:6987–6991
38. Lee MH, Evans MJ, Daemen LL, Sankey OF, Häussermann U (2008) Vibrational property study of $SrGa_2H_2$ and $BaGa_2H_2$ by inelastic neutron scattering and first principles calculations. Inorg Chem 47:1496–1501
39. Lee MH, Björling T, Utsumi T, Moser D, Noréus D, Bull D, Hauback B, Sankey OF, Häussermann U (2008) Crystal structure, electronic structure and vibrational properties of MAlSiH (M = Ca, Sr, Ba): hydrogenation induced semiconductors from AlB_2 type alloys MAlSi. Phys Rev B 78:195209
40. Czybulka A, Pinger B, Schuster HU (1989) New alkaline earth-gallium-silicides, germanides, and stannides with AlB_2-type related structures. Z Anorg Allg Chem 579:151–157
41. Evans MJ, Wu Y, Kranak VF, Newman N, Reller A, Garcia-Garcia FJ, Häussermann U (2009) Structural properties and superconductivity in the ternary intermetallic compounds MAB (M = Ca, Sr, Ba; A = Al, Ga, In; B = Si, Ge, Sn). Phys Rev B 80:064514
42. Giantomassi M, Boeri L, Bachelet GB (2005) Electrons and phonons in the ternary alloy $CaAl_{2-x}Si_x$ as a function of composition. Phys Rev B 72:224512
43. Kuroiwa S, Baron AQR, Muranaka T, Heid R, Bohnen KP, Akimitsu J (2008) Soft-phonon-driven superconductivity in CaAlSi as seen by inelastic scattering. Phys Rev B 77:40503(R)
44. Heid R, Bohnen KP, Renker B, Adelmann P, Wolf T, Ernst D, Schober H (2007) Soft-mode behavior in ternary silicides MAlSi (M = Ca, Sr, Ba). J Low Temp Phys 147:375

45. Imai M, Abe E, Ye JH, Nishida K, Kimura T, Honma K, Abe H, Kitazawa H (2001) Superconductivity of ternary silicide with the AlB2-type structure Sr(Ga0.37, Si0.63)2. Phys Rev Lett 87:077003
46. Lorenz B, Lenzi J, Cmaidalka J, Meng RL, Sun YY, Xue YY, Chu CW (2002) Superconductivity in the C32 intermetallic compounds AAl2-xSix, with A = Ca and Sr, and 0.6 < x < 1.2. Physica C 383:191–196
47. Björling T, Noréus D, Jansson K, Andersson M, Leonova E, Edén M, Hålenius U, Häussermann U (2005) SrAlSiH: a polyanionic semiconductor hydride. Angew Chem Int Ed 44:7269–7273
48. Evans MJ, Holland GP, Häussermann U (2008) Polyanionic gallium hydrides MGaEH from AlB_2 type precursors MGaE (M = Ca, Sr, Ba; E = Si, Ge, Sn). J Am Chem Soc 130:12139–12147
49. Kranak VF, Evans MJ, Daemen LL, Proffen T, Lee MH, Sankey OF, Häussermann U (2009) Structural and dynamic properties of the polyanionic hydrides SrAlGeH and BaAlGeH. Solid State Sci 11:1847–1853
50. Evans MJ, Kranak VF, Holland GP, Daemen LL, Proffen T, Lee MH, Sankey OF, Häussermann U (2009) Structural and dynamic properties of BaInGeH – a rare solid state indium hydride. Inorg Chem 48:5602–5605
51. Evans MJ, Lee MH, Holland GP, Daemen LL, Sankey OF, Häussermann U (2009) Vibrational properties of the gallium monohydrides SrGaGeH, BaGaSiH, BaGaGeH, and BaGaSnH. J Solid State Chem 182:2068–2073
52. Aldridge S, Downs AJ (2001) Hydrides of the main-group metals: new variations on an old theme. Chem Rev 101:3305–3365
53. Ito F, Nakanaga T, Takeo H, Jones H (1994) The infrared spectra of the diatomic hydrides AlH and GaH and intensity analysis from FTIR measurements. J Mol Spectrosc 164:379–389
54. Pullumbi P, Mijoule C, Manceron L, Bouteiller YJ (1994) Aluminum, gallium and indium dihydrides – an IR matrix-isolation and ab-initio study. Chem Phys 185:13–24
55. Pullumbi P, Bouteiller Y, Manceron L, Mijoule C (1994) Aluminum, gallium and indium trihydrides – an IR matrix-isolation and ab-initio study. Chem Phys 185:25–37
56. Irodova AV, Somenkov VA, Bakum SI, Kuznetsova SF (1989) Structure of $NaGaH_4(D_4)$. Z Phys Chem Neue Fol 163:239–242
57. Iniguez J, Yildirim T, Udovic TJ, Sulic M, Jensen CM (2004) Structure and hydrogen dynamics of pure and Ti-doped sodium alanate. Phys Rev B 70:060101(R)

Index

MIX
Papier aus verantwortungsvollen Quellen
Paper from responsible sources
FSC® C105338

If you have any concerns about our products,
you can contact us on
ProductSafety@springernature.com

In case Publisher is established outside the EU,
the EU authorized representative is:
Springer Nature Customer Service Center GmbH
Europaplatz 3, 69115 Heidelberg, Germany

Printed by Libri Plureos GmbH
in Hamburg, Germany